SpringerBriefs in Applied Sciences and Technology

Continuum Mechanics

Series editors

Holm Altenbach, Magdeburg, Germany
Andreas Öchsner, Southport Queensland, Australia

These SpringerBriefs publish concise summaries of cutting-edge research and practical applications on any subject of Continuum Mechanics and Generalized Continua, including the theory of elasticity, heat conduction, thermodynamics, electromagnetic continua, as well as applied mathematics.

SpringerBriefs in Continuum Mechanics are devoted to the publication of fundamentals and applications, presenting concise summaries of cutting-edge research and practical applications across a wide spectrum of fields. Featuring compact volumes of 50 to 125 pages, the series covers a range of content from professional to academic.

More information about this series at http://www.springer.com/series/10528

Sergei Alexandrov

Singular Solutions
in Plasticity

 Springer

Sergei Alexandrov
Institute for Problems in Mechanics
Russian Academy of Sciences
Moscow
Russia

ISSN 2191-530X   ISSN 2191-5318   (electronic)
SpringerBriefs in Applied Sciences and Technology
SpringerBriefs in Continuum Mechanics
ISBN 978-981-10-5226-2   ISBN 978-981-10-5227-9   (eBook)
DOI 10.1007/978-981-10-5227-9

Library of Congress Control Number: 2017944291

Printed on acid-free paper

This Springer imprint is published by Springer Nature
The registered company is Springer Nature Singapore Pte Ltd.
The registered company address is: 152 Beach Road, #21-01/04 Gateway East, Singapore 189721, Singapore

Preface

This monograph concerns with singular solutions in plasticity and their applications. The source of these singular solutions is the maximum friction surface. The presentation of the introductory material and the theoretical developments appear in a text of six chapters. The topics chosen are primarily of interest to engineers as postgraduates and practitioners but they should also serve to capture a readership from among applied mathematicians. The monograph provides both a description of general approaches to finding the asymptotic expansion of solutions in the vicinity of maximum friction surfaces and the specific asymptotic expansions for several rigid plastic material models. Numerical analysis is restricted to plane stain deformation of rigid perfectly plastic material. The potential of singular solutions to describe the generation of narrow fine grain layers in the vicinity of frictional interfaces in metal forming processes is outlined. The possibility of using singular solutions to build up kinematically admissible fields that account for the behavior of real velocity fields near maximum friction surfaces is discussed. Among the topics that are either new or presented in greater detail than would be found in similar texts are the following:

1. Expansions of singular velocity fields near maximum friction surfaces for several rigid plastic models (rigid perfectly plastic material, viscoplastic material and anisotropic material)
2. An approach to using singular velocity fields for describing the generation of fine grain layers in metal forming processes
3. An approach to using singular velocity fields for constructing accurate upper bound solutions.

Chapter 1 concerns with the constitutive equations for several rigid plastic models (rigid perfectly plastic material, viscoplastic material, and anisotropic material). The models for viscoplastic and anisotropic materials reduce to the rigid perfectly plastic model at specific values of input parameters. Several definitions of the maximum friction law are provided. These definitions are connected to the constitutive equations. The constitutive equations introduced in this chapter and the maximum friction law are used in subsequent chapters.

Chapters 2–4 deal with the asymptotic expansions of solutions in the vicinity of maximum friction surfaces for the material models introduced in Chap. 1. First, in Chap. 2, planar and axisymmetric flows of the rigid perfectly plastic material are considered. The main result for the rigid perfectly plastic material model is extended to a class of viscoplastic models in Chap. 3. It is shown that the constitutive equations for viscoplastic material should satisfy certain conditions for the reducibility of viscoplastic solutions to rigid perfectly plastic solutions when the constitutive equations for viscoplastic material reduce to the constitutive equations for rigid perfectly plastic material. Chapter 4 concerns with plane strain deformation of anisotropic materials. A simple analytic solution is provided in each chapter to illustrate the general theory.

In Chap. 5 a numerical method for calculating the strain rate intensity factor, which is the coefficient of the leading singular term in a series expansion of the equivalent strain rate in the vicinity of maximum friction surfaces, is introduced for plane strain deformation of rigid perfectly plastic material. The method is applied to find the strain rate intensity factor in compression of a plastic layer between two parallel plates.

Chapter 6 includes a brief discussion of two applied aspects of the theory of singular solutions. First, it is shown that there is a correlation between the theoretical strain rate intensity factor and the experimental thickness of a fine grain layer generated in the process of direct extrusion of an AZ31 alloy through a conical die. Second, a general kinematically admissible velocity field that accounts for the singular behavior of the real velocity field near maximum friction surfaces is constructed. This kinematically admissible velocity field is then used in conjunction with the upper bound theorem for analysis of forging inside a confined chamber.

Moscow, Russia   Sergei Alexandrov
March 2017

Acknowledgements

This work was made possible by the grant RSCF-14-11-00844.

The figures in Sect. 6.1 (except Fig. 6.1) have been reproduced, courtesy of the publisher of this author's earlier article published in Journal of Manufacturing Science and Engineering, ASME.

Sections 3.1, 4.1, 4.2, Chap. 5, and Sect. 6.2 have been reproduced, courtesy of the publisher of this author's earlier articles, from the following journals:

Meccanica, Springer-Verlag

Continuum Mechanics and Thermodynamics, Springer-Verlag

International Journal of Advanced Manufacturing Technology, Springer-Verlag

Acknowledgements

Contents

Symbols

The intention within the various theoretical developments given in this monograph has been to define each new symbol where it first appears in the text. In this regards each chapter should be treated as self-contained in its symbol content. There are, however, certain symbols that re-appear consistently throughout the text. These symbols are given in the following list.

q, s	Local coordinate system in plane strain and axisymmetry (Figs. 1.1 and 1.2)
u_q, u_s	$q-$ and $s-$ velocity components in plane strain and axisymmetry
D	Strain rate intensity factor
H	Scale factor for q coordinate line (Figs. 1.1 and 1.2)
ξ_{eq}	Equivalent strain rate defined in Eqs. (1.3) and (1.11))
ξ_{qq}, ξ_{ss}	Normal strain rates in (q, s) coordinate system in plane strain and axisymmetry
ξ_{qs}	Shear strain rate in (q, s) coordinate system in plane strain and axisymmetry
$\xi_{\theta\theta}$	Circumferential strain rate
σ	Stress invariant defined in Eq. (1.7)
σ_h	Hydrostatic stress
σ_{qq}, σ_{ss}	Normal stresses in (q, s) coordinate system in plane strain and axisymmetry
σ_{qs}	Shear stress in (q, s) coordinate system in plane strain and axisymmetry
$\sigma_{\theta\theta}$	Circumferential stress
σ_1	Major principal stress
ψ	Orientation of principal stress direction corresponding to σ_1 relative to the q direction (Fig. 1.3)

Chapter 1
Introduction

1.1 Definition of the Subject

Singular solutions considered in the present monograph are restricted to the behavior
of rigid plastic solutions in the vicinity of maximum friction surfaces. The definition for such surfaces depends on the material model chosen. Several definitions are
provided in Sect. 1.4. A common feature of many maximum friction surfaces is that
the equivalent strain rate (the quadratic invariant of the strain rate tensor) approaches
infinity near such surfaces if the regime of sliding occurs. Most likely, the conditions
required by the maximum friction law are never satisfied in real processes. Nevertheless, the asymptotic behavior of singular velocity fields in the vicinity of maximum
friction surfaces can be useful for applications. In particular, the stress intensity factor
is one of basic concepts in linear elastic fracture mechanics. This parameter appears
in asymptotic analyses performed in the vicinity of a sharp crack-tip and is the coefficient of the leading singular term. In spite of the fact that the assumptions under
which the stress intensity factor is determined are not satisfied in real materials (the
crack-tip is not sharp and a region of inelastic deformation exists in its vicinity), this
approach is effective in structural design, for example [24]. In particular, a fracture
criterion is formulated in the form $K = K_c$ where K is the stress intensity factor and
K_c is its critical value, a material parameter. The stress intensity factor controls the
magnitude of stress in the vicinity of crack tips. Analogously, the strain rate intensity
factor introduced in [14] controls the magnitude of the equivalent strain rate in the
vicinity of maximum friction surfaces and the asymptotic behavior of the equivalent
strain rate predicts a very high gradient of this quantity in the vicinity of maximum
friction surfaces. On the other hand, it is well known that plastic strain is one of the
main contributory mechanisms responsible for high gradients in material properties
near frictional interfaces in machining and deformation processes [20]. Therefore, it
is natural to assume, by analogy to linear elastic fracture mechanics, that a new type
of constitutive equations involving the strain rate intensity factor should be developed
to describe the evolution of material properties in the vicinity of surfaces with high
friction in machining and deformation processes [8, 21]. To this end, it is necessary

© The Author(s) 2018
S. Alexandrov, *Singular Solutions in Plasticity*, SpringerBriefs
in Continuum Mechanics, DOI 10.1007/978-981-10-5227-9_1

to have exact asymptotic expansions for the equivalent strain rate near maximum friction surfaces. The present monograph concerns with such expansions for several rigid plastic models such as rigid perfectly plastic solids, viscoplastic solids and anisotropic rigid plastic solids. The elastic component of strain is disregarded and an instantaneous state of stress and velocity is studied. Plane strain and axisymmetric flows are considered.

The approach accepted in this monograph is essentially one developed in [3, 7, 10–14]. Other important results related to the singular behavior of solutions near maximum friction surfaces have been obtained in [4, 9, 15, 19, 35]. In particular, the double shearing model proposed in [36] and further developed in [37] has been considered in [4, 9]. This model is widely used in the mechanics of granular materials.

1.2 Coordinate Systems and Fundamental Equations

Asymptotic analyses near frictional interfaces are facilitated by choosing orthogonal coordinate systems such that one of the coordinate surfaces coincides with the friction surface. Consider an arbitrary smooth curve L in a generic plane of flow in the case of plane strain deformation. This curve represents a friction surface. Introduce a curvilinear orthogonal coordinate system (q, s) assuming that s- lines are straight and the coordinate curve determined by the equation $s = 0$ coincides with L (Fig. 1.1). It is always possible to introduce such a coordinate system in a certain neighborhood of L. The s- axis directed away from the rigid tool and towards the plastic material. Therefore, $s \geq 0$ in the plastic material. It is possible to assume, with no loss of generality, that the scale factor for s- lines is unity. Let σ_{qq}, σ_{ss} and σ_{qs} be the components of the stress tensor in this coordinate system. One of the equilibrium equations is automatically satisfied. The other equilibrium equations are [27]

$$\frac{\partial \sigma_{qq}}{\partial q} + H \frac{\partial \sigma_{qs}}{\partial s} + 2\sigma_{qs} \frac{\partial H}{\partial s} = 0, \quad H \frac{\partial \sigma_{ss}}{\partial s} + \frac{\partial \sigma_{qs}}{\partial q} + (\sigma_{ss} - \sigma_{qq}) \frac{\partial H}{\partial s} = 0. \quad (1.1)$$

Here H is the scale factor for q- lines. The shape of L completely determines H. Let u_q and u_s be the components of velocity relative to the (q, s) coordinates. Then, the non-zero strain rate components are [27]

Fig. 1.1 (q, s) coordinate system in plane strain

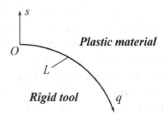

$$\xi_{qq} = \frac{1}{H}\frac{\partial u_q}{\partial q} + \frac{u_s}{H}\frac{\partial H}{\partial s}, \quad \xi_{ss} = \frac{\partial u_s}{\partial s}, \quad 2\xi_{qs} = \frac{\partial u_q}{\partial s} + \frac{1}{H}\frac{\partial u_s}{\partial q} - \frac{u_q}{H}\frac{\partial H}{\partial s}. \quad (1.2)$$

The equivalent strain rate is defined as

$$\xi_{eq} = \sqrt{\frac{2}{3}\left(\xi_{qq}^2 + \xi_{ss}^2 + 2\xi_{qs}^2\right)}. \quad (1.3)$$

The rate of plastic work per unit volume is

$$W = \sigma_{qq}\xi_{qq} + \sigma_{ss}\xi_{ss} + 2\sigma_{qs}\xi_{qs}. \quad (1.4)$$

The physically meaningful solutions must satisfy the inequality

$$\iiint_\Omega W d\Omega < \infty \quad (1.5)$$

where Ω is the volume of the plastic region. The equation of incompressibility that re-appears consistently throughout the text is

$$\frac{\partial u_q}{\partial q} + u_s\frac{\partial H}{\partial s} + H\frac{\partial u_s}{\partial s} = 0. \quad (1.6)$$

Also, it will be convenient to use the stress invariant

$$\sigma = \frac{\sigma_{qq} + \sigma_{ss}}{2}. \quad (1.7)$$

With no loss of generality the friction surface is regarded as motionless. Therefore, the boundary condition

$$u_s = 0 \quad (1.8)$$

for $s = 0$ is always valid.

In the case of axisymmetric flow, L is an arbitrary smooth curve in a generic meridian plane. This curve represents a friction surface. Introduce a cylindrical coordinate system (r, θ, z) and an arbitrary curvilinear orthogonal coordinate system (q, θ, s). The z- axis coincides with the axis of symmetry of flow and the solutions considered in the present monograph are independent of θ. In particular, all derivatives with respect to θ vanish. The circumferential velocity also vanishes. It is assumed that s- lines are straight and that the coordinate curve determined by the equation $s = 0$ coincides with L (Fig. 1.2). It is always possible to introduce such a coordinate system in a certain neighborhood of L. The s-axis directed away from the rigid tool and towards the plastic material. Therefore, $s \geq 0$ in the plastic material. It is possible to assume, with no loss of generality, that the scale factor for s- lines is unity. One of the

Fig. 1.2 (q, s) coordinate
system in generic meridian
plane in axisymmetric flow

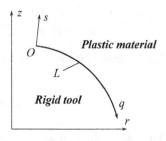

equilibrium equations is automatically satisfied. The other equilibrium equations in
the (q, θ, s) coordinate system are [27]

$$\frac{\partial \sigma_{qq}}{\partial q} + H\frac{\partial \sigma_{qs}}{\partial s} + \frac{(\sigma_{qq} - \sigma_{\theta\theta})}{r}\frac{\partial r}{\partial q} + \sigma_{qs}\left(2\frac{\partial H}{\partial s} + \frac{H}{r}\frac{\partial r}{\partial s}\right) = 0, \qquad (1.9)$$

$$\frac{\partial \sigma_{ss}}{\partial s} + \frac{\partial \sigma_{qs}}{H\partial q} + \frac{(\sigma_{ss} - \sigma_{qq})}{H}\frac{\partial H}{\partial s} + \frac{(\sigma_{ss} - \sigma_{\theta\theta})}{r}\frac{\partial r}{\partial s} = 0.$$

As before, H is the scale factor for q- lines. $\sigma_{\theta\theta}$ is the circumferential stress. The
non-zero strain rate components in the (q, θ, s) coordinate system are [27]

$$\xi_{qq} = \frac{1}{H}\frac{\partial u_q}{\partial q} + \frac{u_s}{H}\frac{\partial H}{\partial s}, \quad \xi_{ss} = \frac{\partial u_s}{\partial s}, \quad \xi_{\theta\theta} = \frac{u_r}{r}, \qquad (1.10)$$

$$2\xi_{qs} = \frac{\partial u_q}{\partial s} + \frac{1}{H}\frac{\partial u_s}{\partial q} - \frac{u_q}{H}\frac{\partial H}{\partial s}$$

where u_r is the radial velocity. The boundary condition (1.8) is valid. The equivalent
strain rate is defined as

$$\xi_{eq} = \sqrt{\frac{2}{3}\left(\xi_{qq}^2 + \xi_{ss}^2 + \xi_{\theta\theta}^2 + 2\xi_{qs}^2\right)}. \qquad (1.11)$$

The rate of plastic work per unit volume is

$$W = \sigma_{qq}\xi_{qq} + \sigma_{\theta\theta}\xi_{\theta\theta} + \sigma_{ss}\xi_{ss} + 2\sigma_{qs}\xi_{qs}. \qquad (1.12)$$

The physically meaningful solutions must satisfy the inequality (1.5). The equation
of incompressibity that re-appears consistently throughout the text is

$$\frac{\partial u_q}{\partial q} + u_s\frac{\partial H}{\partial s} + H\frac{\partial u_s}{\partial s} + H\frac{u_r}{r} = 0. \qquad (1.13)$$

Fig. 1.3 Orientation of characteristics based coordinates and σ_1 relative to the (q, s) coordinate system

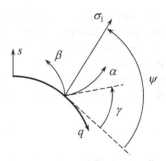

Also, it will be convenient to use the stress invariant

$$\sigma_h = \frac{\sigma_{qq} + \sigma_{\theta\theta} + \sigma_{ss}}{3}. \tag{1.14}$$

This is the hydrostatic stress.

The subsequent analyses are based on the following assumptions:

Assumption 1.1

(i)   stresses and velocities are bounded everywhere;
(ii) derivatives of stress and velocity components with respect to q are bounded everywhere;
(iii) solutions are represented by power series of s in the vicinity of $s = 0$.

In the case of hyperbolic equations, it is sometimes convenient to use characteristics based coordinate systems. In the present monograph, the use of such coordinate systems is restricted to plane strain deformation. Moreover, in all cases considered the characteristics for the stresses and the velocities coincide and, therefore, there are only two distinct characteristic directions at a point. The characteristic directions are orthogonal. Let the two families of characteristics be labeled by the parameters α and β. The α- and β- lines are regarded as a pair of right-handed curvilinear orthogonal coordinate curves. The anti-clockwise angular rotation of the α- line from the q-line of the (q, s) coordinate system is denoted by γ (Fig. 1.3). Let σ_1 and σ_2 be the principal stresses in planes of flow. It is possible to assume with no loss of generality that $\sigma_1 > \sigma_2$. The anti-clockwise angular rotation of the principal stress direction corresponding to σ_1 from the q- line of the (q, s) coordinate system is denoted by ψ (Fig. 1.3).

1.3   Material Models

The simplest rigid plastic model is a rigid perfectly plastic solid. A great account on this model is given in [22]. The constitutive equations of many rigid plastic models, including the models considered in the present monograph, reduce to the constitutive equations of the rigid perfectly plastic solid at specific values of input parameters. Nevertheless, the solution behavior in the vicinity of maximum friction surfaces essentially depends on the model chosen, independently of how close the input parameters are to these specific values. The constitutive equations of the rigid perfectly plastic solid are a pressure-independent yield criterion and its associated flow rule. In the case of plane strain deformation, all possible yield criteria reduce to

$$\left(\sigma_{qq} - \sigma_{ss}\right)^2 + 4\sigma_{qs}^2 = 4k_0^2 \tag{1.15}$$

where k_0 is the shear yield stress, a material constant. The flow rule associated with the yield criterion (1.15) is

$$\xi_{qq} = \lambda \left(\sigma_{qq} - \sigma_{ss}\right), \quad \xi_{ss} = \lambda \left(\sigma_{ss} - \sigma_{qq}\right), \quad \xi_{qs} = 2\lambda\sigma_{qs}. \tag{1.16}$$

Here λ is a non-negative multiplier. It is seen from these equations and Eq. (1.2) that Eq. (1.6) is satisfied. The system of equations consisting of Eqs. (1.1), (1.2), (1.15), and (1.16) is hyperbolic. The angle between the σ_1- direction and the characteristic directions is equal to $\pi/4$. In particular (Fig. 1.3),

$$\psi - \gamma = \frac{\pi}{4}. \tag{1.17}$$

In the case of axisymmetric deformation, the subsequent analysis is restricted to the von Mises yield criterion. In the case under consideration this criterion reads

$$\left(\sigma_{qq} - \sigma_{ss}\right)^2 + (\sigma_{ss} - \sigma_{\theta\theta})^2 + \left(\sigma_{\theta\theta} - \sigma_{qq}\right)^2 + 6\sigma_{qs}^2 = 6k_0^2. \tag{1.18}$$

The flow rule associated with the yield criterion (1.18) is

$$\xi_{qq} = \lambda \left(2\sigma_{qq} - \sigma_{ss} - \sigma_{\theta\theta}\right), \quad \xi_{ss} = \lambda \left(2\sigma_{ss} - \sigma_{\theta\theta} - \sigma_{qq}\right), \tag{1.19}$$
$$\xi_{\theta\theta} = \lambda \left(2\sigma_{\theta\theta} - \sigma_{ss} - \sigma_{qq}\right), \quad \xi_{qs} = 3\lambda\sigma_{qs}.$$

It is seen from these equations and Eq. (1.10) that Eq. (1.13) is satisfied.

A wide class of viscoplastic models is obtained by assuming that the shear yield stress depends on the equivalent strain rate. An overview of viscoplastic models is given in [16]. The yield criteria (1.15) and (1.18) become

$$\left(\sigma_{qq} - \sigma_{ss}\right)^2 + 4\sigma_{qs}^2 = 4k_0^2 f^2 \left(\xi_{eq}\right) \tag{1.20}$$

and

$$\left(\sigma_{qq} - \sigma_{ss}\right)^2 + (\sigma_{ss} - \sigma_{\theta\theta})^2 + \left(\sigma_{\theta\theta} - \sigma_{qq}\right)^2 + 6\sigma_{qs}^2 = 6k_0^2 f^2 \left(\xi_{eq}\right), \tag{1.21}$$

respectively. In Eqs. (1.20) and (1.21), $f\left(\xi_{eq}\right)$ is an arbitrary function of the equivalent strain rate satisfying the condition $df/d\xi_{eq} \geq 0$ for all ξ_{eq}. It is evident that Eqs. (1.20) and (1.21) reduce to Eqs. (1.15) and (1.18), respectively, if $f\left(\xi_{eq}\right) \equiv 1$. The flow rules (1.16) and (1.19) are valid. In the terminology of Oldroyd [30], the non-Newtonian liquids are defined by the condition $f(0) = 0$ and the non-Newtonian solids by the condition $f(0) = 1$. This classification is not important for the asymptotic behavior of solutions in the vicinity of maximum friction surfaces. Of great importance is the behavior of the function $f\left(\xi_{eq}\right)$ as $\xi_{eq} \to \infty$. In particular, all viscoplastic models can be conveniently divided into two groups. The models of the first group are defined by the condition

$$f\left(\xi_{eq}\right) \to \infty \quad \text{as} \quad \xi_{eq} \to \infty \tag{1.22}$$

and the models of the other group by the condition

$$f\left(\xi_{eq}\right) \to k_s/k_0 < \infty \quad \text{as} \quad \xi_{eq} \to \infty. \tag{1.23}$$

Here k_s is the saturation stress and $k_s > k_0$. By assumption k_s is a material constant.

It has been shown in [32] that the plane strain yield criterion of any incompressible anisotropic material which complies with the principle of maximum plastic work is expressed solely in terms of the stress variables

$$t = \left(\sigma_{qq} - \sigma_{ss}\right)/2 \quad \text{and} \quad \tau = \sigma_{qs}. \tag{1.24}$$

Therefore, the yield criterion for such materials can be written as

$$F\left(t, \tau\right) = 0. \tag{1.25}$$

The function $F\left(t, \tau\right)$ must satisfy the standard requirements imposed on the yield criteria in the theory of rigid plastic materials based on the associated flow rule, but which is otherwise arbitrary. The system of equations consisting of Eqs. (1.1), (1.2), (1.25) and the associated flow rule is hyperbolic. Therefore, the characteristics based coordinate system (α, β) can be introduced as shown in Fig. 1.3. The yield contour corresponding to Eq. (1.25) is illustrated in the Mohr stress plane in Fig. 1.4. It is worthy of note that the angle between the outward normal to the contour and τ-axis measured from the axis anticlockwise at some point M_* is equal to 2γ where γ is defined as shown in Fig. 1.3 [32]. A consequence of the associated flow rule is [32]

$$\xi_{qq} + \xi_{ss} = 0, \quad \left(\xi_{qq} - \xi_{ss}\right)\cos 2\gamma + \xi_{qs}\sin 2\gamma = 0. \tag{1.26}$$

The first equation is equivalent to Eq. (1.6).

Fig. 1.4 Yield contour in
Mohr plane

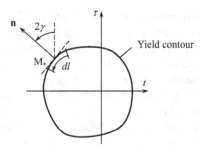

1.4   Maximum Friction Law

The general friction law includes two regimes, the regime of sticking and the regime of sliding. Throughout the present monograph the regime of sliding is always assumed. Moreover, it is always possible to assume, with no loss of generality, that

$$\sigma_{qs} > 0 \tag{1.27}$$

in the vicinity of maximum friction surfaces in the coordinate systems shown in Figs. 1.1 and 1.2.

It is seen from Eqs. (1.15) and (1.18) that $\sigma_{qs} \leq k_0$ in the case of the rigid perfectly plastic model. Therefore, in this case it is natural to define the maximum friction law as $\tau_f = k_0$ where τ_f is the friction stress. Taking into account Eq. (1.27) this equation becomes

$$\sigma_{qs} = k_0 \tag{1.28}$$

at the friction surface. This boundary condition is often adopted at the tool—chip interface (at least, over a portion of this interface) in machining processes [1, 2, 18, 26, 28, 29, 31, 40]. This zone is usually called the sticking zone. The existence of such zones has been reported in deformation processes as well [25, 33, 34, 39]. In the metal forming literature, it is often assumed that the maximum friction law represents sticking friction conditions [17, 38]. However, it is worthy of note that the actual seizure (the velocity vector is continuous across the interface) may or may not occur in theoretical solutions. It depends on the constitutive equations chosen and other boundary conditions. Also, the interpretation of experimental results on this issue is controversial [23].

In the case of the rigid perfectly plastic model, an essential difference between plane strain and axisymmetric flows is that the plane strain equations are hyperbolic whereas the axisymmetric equations are not. In the case of plane strain flow it is possible to provide another formulation of the maximum friction law using properties of characteristic curves. In particular, the magnitude of the shear stress along characteristic curves is equal to k_0 [22]. Therefore, the friction surface coincides with a characteristic or an envelope of characteristics. The characteristic curves are

Fig. 1.5 Principal stress and
characteristic directions in
rigid perfectly plastic
material

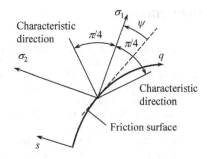

orthogonal and the characteristic directions make the angles $\pm\pi/4$ with the direction
of σ_1 (Fig. 1.5). Therefore, taking into account that $\sigma_1 > \sigma_2$ and using Eq. (1.27) it is
possible to conclude that the boundary condition (1.28) is equivalent to

$$\psi = \pi/4. \tag{1.29}$$

Using Eq. (1.17) the boundary condition (1.29) can be represented as

$$\gamma = 0 \tag{1.30}$$

at the friction surface.

In the case of viscoplastic models which satisfy the condition (1.22) the maximum
shear stress may increase with no limit. Therefore, sliding cannot occur if the friction
stress is assumed to be equal to the local shear yield stress [5, 6]. This type of
viscoplastic models is not considered in the present monograph.

In the case of viscoplastic models which satisfy the condition (1.23) the maximum
friction law reads

$$\sigma_{qs} = k_s. \tag{1.31}$$

This law generalizes the maximum friction law (1.28).

The model of anisotropic plasticity introduced in [32] is hyperbolic. Therefore,
it is most convenient to formulate the maximum friction law in terms of the angle
between the friction surface and one of the characteristic directions. It is seen from
Fig. 1.3 that a necessary condition that an α- line or an envelope of β- lines coincides
with the friction surface is

$$\gamma = 0. \tag{1.32}$$

It is understood here that γ is to be calculated at the friction surface. Equation (1.32)
is the maximum friction law for the model of anisotropic plasticity under consideration. It is evident that the maximum friction law for isotropic rigid perfectly plastic
materials in the form of Eq. (1.30) coincides with the maximum friction law for
anisotropic materials in the form of Eq. (1.32).

References

1. Agmell M, Ahadi A, Stahl J-E (2014) Identification of plastic constants from orthogonal cutting and inverse analysis. Mech Mater 77:43–51
2. Akbar F, Mativenga PT, Sheikh MA (2010) An experimental and coupled thermo-mechanical finite element study of heat partition effects in machining. Int J Adv Manuf Technol 46:491–507
3. Alexandrov S (1992) Discontinuous velocity fields due to arbitrary strains in an ideal rigid-plastic body. Sov Phys Dokl 37:283–284
4. Alexandrov S (2005) Singular solutions in an axisymmetric flow of a medium obeying the double shear model. J Appl Mech Techn Phys 46:766–771
5. Alexandrov S, Alexandrova N (2000) On the maximum friction law in viscoplasticity. Mech Time Depend Mater 4:99–104
6. Alexandrov S, Danilov V, Chikanova N (2000) On the stagnation zone in a simulation of axisymmetric pressure metal forming under creep. Mech Solids 35:127–129
7. Alexandrov S, Druyanov B (1992) Friction conditions for plastic bodies. Mech Solids 27:110–115
8. Alexandrov SE, Goldstein RV (2015) On constructing constitutive equations in metal thin layer near friction surfaces in material forming processes. Dokl Phys 60:39–41
9. Alexandrov S, Lyamina E (2002) Singular solutions for plane plastic flow of pressure-dependent materials. Dokl Phys 47:308–311
10. Alexandrov S, Mishuris G (2007) Viscoplasticity with a saturation stress: distinguished features of the model. Arch Appl Mech 77:35–47
11. Alexandrov S, Mishuris G (2009) Qualitative behaviour of viscoplastic solutions in the vicinity of maximum-friction surfaces. J Eng Math 65:143–156
12. Alexandrov S, Mustafa Y (2013) Singular solutions in viscoplasticity under plane strain conditions. Meccanica 48:2203–2208
13. Alexandrov S, Mustafa Y (2015) Quasi-static axially symmetric viscoplastic flows near very rough walls. Appl Math Model 39:4599–4606
14. Alexandrov S, Richmond O (2001) Singular plastic flow fields near surfaces of maximum friction stress. Int J Non Linear Mech 36:1–11
15. Atkins AG, Rowe GM, Johnson W (1982) Shear strains and strain-rates in kinematically admissible velocity fields. Int J Mech Eng Educ 10:265–278
16. Barnes HA (1999) The yield stress a review or $\pi\alpha\nu\tau\alpha$ $\rho\varepsilon\iota$ everything flows. J Non Newton Fluid Mech 81:133–178
17. Chandrasekharan S, Palaniswamy H, Jain N, Ngaile G, Altan T (2005) Evaluation of stamping lubricants at various temperature levels using the ironing test. Int J Mach Tools Manuf 45:379–388
18. Chen G, Li J, He Y, Ren C (2014) A new approach to the determination of plastic flow stress and failure initiation strain for aluminium alloys cutting process. Comp Mater Sci 95:568–578
19. Durban D (1999) Friction and singularities in steady penetration. In: Durban D, Pearson JRA (eds) IUTAM symposium on non-linear singularities in deformation and flow. Kluwer, pp 141–154
20. Griffiths BJ (1987) Mechanisms of white layer generation with reference to machining and deformation processes. ASME J Trib 109:525–530
21. Goldstein RV, Alexandrov SE (2015) An approach to prediction of microstructure formation near friction surfaces at large plastic strains. Phys Mesomech 18:223–227
22. Hill R (1950) The mathematical theory of plasticity. Clarendon Press, Oxford
23. Jaspers SPFC, Dautzenberg JH (2002) Material behaviour in metal cutting: strains, strain rates and temperature in chip of formation. J Mater Process Technol 121:123–135
24. Kanninen MF, Popelar CH (1985) Advanced fracture mechanics. Oxford University Press, New York
25. Kim T-K, Ikeda K (2000) Flow behavior of the billet surface layer in porthole die extrusion of aluminium. Metall Mater Trans 31A:1635–1643

26. Lalwani DI, Mehta NK, Jain PK (2009) Extension of Oxleys predictive machining theory for Johnson and Cook flow stress model. J Mater Process Technol 209:5305–5312
27. Malvern LE (1969) Introduction to the mechanics of a continuous medium. Prentice-Hall, Englewood Cliffs
28. Molinari A, Cheriguene R, Miguelez H (2012) Contact variables and thermal effects at the tool-chip interface in orthogonal cutting. Int J Solids Struct 49:3774–3796
29. Ng E-G, Aspinwall DK, Brazil D, Monaghan J (1999) Modelling of temperature and forces when orthogonally machining hardened steel. Int J Mach Tool Manuf 39:885–903
30. Oldroyd JG (1956) Non-Newtonian flow of liquids and solids. In: Eirich FR (ed) Rheology: theory and applications, vol 1. Academic Press, New York, pp 653–682
31. Ramesh A, Melkote SN (2008) Modeling of white layer formation under thermally dominate condition in orthogonal machining of hardened AISI 52100 steel. Int J Mach Tool Manuf 48:402–414
32. Rice JR (1973) Plane strain slip line theory for anisotropic rigid/plastic materials. J Mech Phys Solids 21:63–74
33. Sanabria V, Mueller S, Gall S, Reimers W (2014) Investigation of friction boundary conditions during extrusion of aluminium and magnesium alloys. Key Eng Mater 611–612:997–1004
34. Sanabria V, Mueller S, Reimers W (2014) Microstructure evolution of friction boundary layer during extrusion of AA 6060. Proc Eng 81:586–591
35. Sokolovskii VV (1956) Equations of plastic flow in surface layer. Prikl Math Mech 20:328–334 [in Russian]
36. Spencer AJM (1964) A theory of the kinematics of ideal soils under plane strain conditions. J Mech Phys Solids 12:337–351
37. Spencer AJM (1982) Deformation of ideal granular materials. In: Hopkins HG, Sewell MJ (eds) Mechanics of solids. The Rodney Hill 60th anniversary volume. Pergamon Press, Oxford, pp 607–652
38. Wang J-P (2001) A new evaluation to friction analysis for the ring test. Int J Mach Tools Manuf 41:311–324
39. Widerøe F, Welo T (2012) Conditions for sticking friction between aluminium alloy AA6060 and tool steel in hot forming. Key Eng Mater 491:121–128
40. Yiğit K, Tuğrul Ö (2006) Predictive analytical and thermal modeling of orthogonal cutting process Part I: Predictions of tool forces, stresses, and temperature distributions. ASME J Manuf Sci Eng 128:435–444

Chapter 2
Rigid Perfectly Plastic Material

2.1 Plane Strain Deformation

The (q, s) coordinate system illustrated in Fig. 1.1 will be used. The system of equations consisting of Eqs. (1.1), (1.2), (1.15), and (1.16) is hyperbolic [3]. Therefore, the maximum friction law in the form of Eq. (1.29) may be adopted. The yield criterion (1.15) is satisfied by the following standard substitution [3]

$$\sigma_{qq} = \sigma + k_0 \cos 2\psi, \quad \sigma_{ss} = \sigma - k_0 \cos 2\psi, \quad \sigma_{qs} = k_0 \sin 2\psi. \tag{2.1}$$

Here σ is the stress invariant introduced in Eq. (1.7). It is worthy of note that σ is equal to the hydrostatic stress for the model under consideration. Substituting Eq. (2.1) into Eqs. (1.1) and (1.16) results in

$$\frac{\partial \sigma}{\partial q} - 2k_0 \sin 2\psi \frac{\partial \psi}{\partial q} + 2Hk_0 \cos 2\psi \frac{\partial \psi}{\partial s} + 2k_0 \sin 2\psi \frac{\partial H}{\partial s} = 0, \tag{2.2}$$

$$H \frac{\partial \sigma}{\partial s} + 2Hk_0 \sin 2\psi \frac{\partial \psi}{\partial s} + 2k_0 \cos 2\psi \frac{\partial \psi}{\partial q} - 2k_0 \cos 2\psi \frac{\partial H}{\partial s} = 0.$$

and

$$\xi_{qq} = \lambda_1 \cos 2\psi, \quad \xi_{ss} = -\lambda_1 \cos 2\psi, \quad \xi_{qs} = \lambda_1 \sin 2\psi, \tag{2.3}$$

respectively. In Eq. (2.3), $\lambda_1 = 2k_0\lambda > 0$. It is seen from Eqs. (1.29) and (2.3) that $\xi_{qq} = \xi_{ss} = 0$ along the friction surface unless

$$\lambda_1 \to \infty \tag{2.4}$$

as $s \to 0$. The condition $\xi_{qq} = 0$ means that $s = 0$ is a characteristic curve [3] (i.e. the friction surface coincides with this characteristic curve). In this case the solution

© The Author(s) 2018
S. Alexandrov, *Singular Solutions in Plasticity*, SpringerBriefs
in Continuum Mechanics, DOI 10.1007/978-981-10-5227-9_2

is not singular. Therefore, this case is not considered in the present monograph. It is assumed that

$$\xi_{qq} \neq 0 \tag{2.5}$$

along the friction surface and Eq. (2.4) is satisfied. This means that the friction surface coincides with an envelope of characteristics. It follows from Eqs. (1.27), (1.29), (2.3) and (2.4) that

$$\xi_{qs} \to \infty \tag{2.6}$$

as $s \to 0$. Using assumption (ii) of Assumptions 1.1 (see p. 5) it is possible to conclude that the derivative $\partial u_s / \partial q$ is bounded at $s = 0$. Then, it follows from Eqs. (1.2) and (2.6) that

$$\frac{\partial u_q}{\partial s} \to \infty \tag{2.7}$$

as $s \to 0$. Using assumption (iii) of Assumptions 1.1 (see p. 5) it is possible to represent the angle ψ as

$$\psi = \frac{\pi}{4} + \psi_0 s^\beta + o\left(s^\beta\right) \tag{2.8}$$

as $s \to 0$. Here ψ_0 is independent of s and

$$\beta > 0. \tag{2.9}$$

It follows from Eq. (2.8) that

$$\sin 2\psi = 1 - 2\psi_0^2 s^{2\beta} + o\left(s^{2\beta}\right), \quad \cos 2\psi = -2\psi_0 s^\beta + o\left(s^\beta\right) \tag{2.10}$$

as $s \to 0$. Substituting Eq. (2.10) into Eq. (2.3) gives

$$\xi_{qq} = -\xi_{ss} = -2\psi_0 \lambda_1 s^\beta + o\left(s^\beta\right) \tag{2.11}$$

as $s \to 0$. Using assumption (ii) of Assumptions 1.1 (see p. 5) and Eq. (1.2) it is possible to conclude that ξ_{qq} is bounded at $s = 0$. Therefore, it follows from Eqs. (2.5) and (2.11) that

$$\lambda_1 = \lambda_0 s^{-\beta} + o\left(s^{-\beta}\right) \tag{2.12}$$

as $s \to 0$. Here λ_0 is independent of s. Substituting Eq. (2.12) into Eq. (2.3) for ξ_{qs} and taking into account Eq. (2.10) results in

$$\xi_{qs} = \lambda_0 s^{-\beta} + o\left(s^{-\beta}\right) \tag{2.13}$$

as $s \to 0$. Since $\sigma_{qs} \neq 0$ and the strain rate components ξ_{qq} and ξ_{ss} are bounded at $s = 0$, it follows from Eqs. (1.4) and (2.13) that $O(W) = O(\xi_{qs}) = O(s^{-\beta})$

as $s \to 0$. Therefore, the inequality (1.5) is satisfied if $\beta < 1$. This inequality and Eq. (2.9) combine to give

$$0 < \beta < 1. \tag{2.14}$$

Substituting Eqs. (2.8) and (2.10) into Eq. (2.2)[1] leads to

$$\left[\frac{\partial \sigma}{\partial q} + 2k_0 \frac{\partial H}{\partial s} + o\,(1) \right] - \left[2k_0 \frac{d\psi_0}{dq} s^\beta + o\,\left(s^\beta\right) \right] - \\ \left\{ 4k_0 H \psi_0^2 \beta s^{(2\beta - 1)} + o\,\left[s^{(2\beta - 1)}\right] \right\} = 0 \tag{2.15}$$

as $s \to 0$. In this equation, H and $\partial H / \partial s$ are understood to be calculated at $s = 0$. Using assumption (ii) of Assumptions 1.1 (see p. 5) it is possible to conclude that the derivative $\partial \sigma / \partial q$ is bounded at $s = 0$. Therefore, it follows from Eq. (2.15) that it is necessary to examine two cases, namely $2\beta - 1 = \beta$ and $2\beta - 1 = 0$. Assume that $2\beta - 1 = \beta$. Then, $\beta = 1$. This contradicts Eq. (2.14). Therefore, $2\beta - 1 = 0$ or

$$\beta = \frac{1}{2}. \tag{2.16}$$

Then, it follows from Eq. (2.15) that

$$\sigma = \Sigma_0 + \Sigma_1 \sqrt{s} + o\,\left(\sqrt{s}\right) \tag{2.17}$$

as $s \to 0$. Here Σ_0 and Σ_1 are independent of s. Substituting Eqs. (2.8), (2.10), (2.16) and (2.17) into Eq. (2.2)[2] yields

$$\frac{H \Sigma_1}{2} s^{-1/2} + H k_0 \psi_0 s^{-1/2} + o\,\left(s^{-1/2}\right) = 0$$

as $s \to 0$. Therefore,

$$\Sigma_1 = -2k_0 \psi_0. \tag{2.18}$$

Substituting this equation into Eq. (2.17) leads to

$$\sigma = \Sigma_0 - 2k_0 \psi_0 \sqrt{s} + o\,\left(\sqrt{s}\right) \tag{2.19}$$

as $s \to 0$. Eliminating σ in Eq. (2.15) by means of Eq. (2.19) shows that Eq. (2.15) contains the term $-4k_0\,(d\psi_0/dq)\,\sqrt{s}$. This is the only term of the order $\sqrt{s}$ as $s \to 0$ involved in this equation. Therefore, it is necessary to assume that ψ_0 is constant. Since the strain components ξ_{qq} and ξ_{ss} are bounded at $s = 0$, it follows from Eqs. (1.3), (2.13) and (2.16) that

$$\xi_{eq} = \frac{D}{\sqrt{s}} + o\left(\frac{1}{\sqrt{s}}\right) \tag{2.20}$$

as $s \to 0$. Here D is the strain rate intensity factor [2]. The distribution of stresses near the maximum friction surface is found from Eqs. (2.1), (2.10), (2.16), and (2.19) as

$$\sigma_{qq} = \Sigma_0 - 4k_0\psi_0\sqrt{s} + o\left(\sqrt{s}\right),$$
$$\sigma_{ss} = \Sigma_0 + o\left(\sqrt{s}\right), \quad \sigma_{qs} = k_0\left(1 - 2\psi_0^2 s\right) + o\left(s\right) \tag{2.21}$$

as $s \to 0$. The distribution of velocity follows from Eqs. (2.11) and (2.13). In particular, substituting Eq. (2.11) for ξ_{ss} into Eq. (1.2), eliminating λ_1 by means of Eq. (2.12) and integrating result in

$$u_s = 2\psi_0\lambda_0 s + o\left(s\right) \tag{2.22}$$

as $s \to 0$. Here it has been taken into account that this velocity component should satisfy the boundary condition (1.8). Taking into account Eqs. (2.7), (2.13) and (2.16) the equation for the shear strain rate in Eq. (1.2) can be represented as $\partial u_q/\partial s = 2\lambda_0 s^{-1/2} + o\left(s^{-1/2}\right)$ as $s \to 0$. Integrating leads to

$$u_q = u_0 + 4\lambda_0\sqrt{s} + o\left(\sqrt{s}\right) \tag{2.23}$$

as $s \to 0$. Here u_0 is independent of s.

2.2   Axisymmeric Deformation

The (q, θ, s) coordinate system illustrated in Fig. 1.2 will be used. The system of equations consisting of Eqs. (1.9), (1.10), (1.18), and (1.19) is not hyperbolic [3]. Therefore, the maximum friction law in the form of Eq. (1.28) should be adopted. Equations (1.14), (1.18) and (1.28) combine to give

$$\sigma_{qq} = \sigma_{ss} = \sigma_{\theta\theta} = \sigma_h \tag{2.24}$$

at $s = 0$. Substituting this equation into Eq. (1.19) for $\xi_{\theta\theta}$ shows that $\xi_{\theta\theta} = 0$ unless

$$\lambda \to \infty \tag{2.25}$$

as $s \to 0$. The condition $\xi_{\theta\theta} = 0$ contradicts Eq. (1.10) for $\xi_{\theta\theta}$ if $u_r \neq 0$ and $u_r \neq 0$ if the regime of sliding occurs (the velocity vector is not equal to zero) and the tangent to the friction surface is not parallel to the z- axis, which is the axis of symmetry of the process, at a generic point M of the friction surface (Fig. 2.1). Therefore, Eq. (2.25) is in general valid. Then, it follows from Eqs. (1.19) and (1.28) that

$$\xi_{qs} \to \infty \tag{2.26}$$

Fig. 2.1 Illustration of the condition $\xi_{\theta\theta} \neq 0$ at friction surface

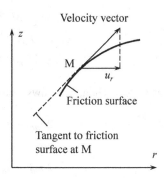

as $s \to 0$. Using assumption (ii) of Assumptions 1.1 (see p. 5) it is possible to conclude that the derivative $\partial u_s / \partial q$ is bounded at $s = 0$. Then, it follows from Eqs. (1.10) and (2.26) that

$$\frac{\partial u_q}{\partial s} \to \infty \tag{2.27}$$

as $s \to 0$. Using assumption (iii) of Assumptions 1.1 (see p. 5) it is possible to represent u_q as

$$u_q = u_0 + u_1 s^\beta + o\left(s^\beta\right) \tag{2.28}$$

as $s \to 0$. Here u_0 and u_1 are independent of s. It follows from assumption (i) of Assumptions 1.1 (see p. 5) that $\beta > 0$. On the other hand, it is seen from Eqs. (2.27) and (2.28) that $\beta < 1$. Hence,

$$0 < \beta < 1. \tag{2.29}$$

Using assumptions (i) and (ii) of Assumptions 1.1 (see p. 5) it is possible to find from Eq. (1.10) that the strain rate components ξ_{qq} and $\xi_{\theta\theta}$ are bounded as $s \to 0$. Then, it follows from Eqs. (1.10) and (1.13) that the strain rate component ξ_{ss} is bounded as $s \to 0$. Therefore,

$$\xi_{qq} = O\left(1\right), \quad \xi_{ss} = O\left(1\right) \quad \text{and} \quad \xi_{\theta\theta} = O\left(1\right) \tag{2.30}$$

as $s \to 0$. Substituting Eq. (2.28) into Eq. (1.10) leads to

$$\xi_{qs} = O\left(s^{\beta-1}\right) \tag{2.31}$$

as $s \to 0$. It is seen from Eqs. (1.12), (2.29), (2.30) and (2.31) that the inequality (1.5) is satisfied. Equations (1.19) for ξ_{qs}, (1.28) and (2.31) combine to give $\lambda = O\left(s^{\beta-1}\right)$ as $s \to 0$. This equation can be rewritten as

$$\lambda = \lambda_0 s^{\beta-1} + o\left(s^{\beta-1}\right) \tag{2.32}$$

as $s \to 0$. Here λ_0 is independent of s. Substituting Eq. (2.32) into Eq. (1.19) leads to

$$\xi_{qq} = \left[\lambda_0 s^{\beta-1} + o\left(s^{\beta-1}\right)\right]\left(2\sigma_{qq} - \sigma_{ss} - \sigma_{\theta\theta}\right),$$
$$\xi_{ss} = \left[\lambda_0 s^{\beta-1} + o\left(s^{\beta-1}\right)\right]\left(2\sigma_{ss} - \sigma_{\theta\theta} - \sigma_{qq}\right), \tag{2.33}$$
$$\xi_{\theta\theta} = \left[\lambda_0 s^{\beta-1} + o\left(s^{\beta-1}\right)\right]\left(2\sigma_{\theta\theta} - \sigma_{ss} - \sigma_{qq}\right).$$

as $s \to 0$. Then, it follows from Eq. (2.30) that

$$2\sigma_{qq} - \sigma_{ss} - \sigma_{\theta\theta} = A_q s^{1-\beta} + o\left(s^{1-\beta}\right),$$
$$2\sigma_{ss} - \sigma_{\theta\theta} - \sigma_{qq} = A_s s^{1-\beta} + o\left(s^{1-\beta}\right), \tag{2.34}$$
$$2\sigma_{\theta\theta} - \sigma_{qq} - \sigma_{ss} = A_\theta s^{1-\beta} + o\left(s^{1-\beta}\right)$$

as $s \to 0$. Here A_q, A_s and A_θ are independent of s. Equation (2.34) can be transformed to

$$\sigma_{qq} - \sigma_{ss} = \frac{\left(A_q - A_s\right)}{3} s^{1-\beta} + o\left(s^{1-\beta}\right),$$
$$\sigma_{ss} - \sigma_{\theta\theta} = \frac{\left(A_s - A_\theta\right)}{3} s^{1-\beta} + o\left(s^{1-\beta}\right), \tag{2.35}$$
$$\sigma_{\theta\theta} - \sigma_{qq} = \frac{\left(A_\theta - A_q\right)}{3} s^{1-\beta} + o\left(s^{1-\beta}\right)$$

as $s \to 0$. Using assumption (iii) of Assumptions 1.1 (see p. 5) and taking into account Eq. (1.28) it is possible to represent σ_{qs} as

$$\sigma_{qs} = k_0 + O\left(s^\omega\right) \tag{2.36}$$

as $s \to 0$. Here $\omega > 0$. Substituting Eqs. (2.35) and (2.36) into Eq. (1.18) results in

$$\frac{\left[\left(A_q - A_s\right)^2 + \left(A_s - A_\theta\right)^2 + \left(A_\theta - A_q\right)^2\right]}{9} s^{2(1-\beta)} = O\left(s^\omega\right). \tag{2.37}$$

as $s \to 0$. It is worthy of note that the coefficient of $s^{2(1-\beta)}$ never vanishes. Then, it follows from Eq. (2.37) that $\omega = 2(1 - \beta)$ and Eq. (2.36) can be rewritten as

$$\sigma_{qs} = k_0 + k_1 s^{2(1-\beta)} + o\left[s^{2(1-\beta)}\right]. \tag{2.38}$$

as $s \to 0$. Here k_1 is independent of s. Using Eq. (1.14) it is possible to rewrite Eq. (2.34) as

$$\sigma_{qq} - \sigma_h = \tfrac{A_q}{3} s^{1-\beta} + o\left(s^{1-\beta}\right),$$
$$\sigma_{ss} - \sigma_h = \tfrac{A_s}{3} s^{1-\beta} + o\left(s^{1-\beta}\right), \quad \sigma_{\theta\theta} - \sigma_h = \tfrac{A_\theta}{3} s^{1-\beta} + o\left(s^{1-\beta}\right) \tag{2.39}$$

as $s \to 0$. The derivative $\partial \sigma_{ss} / \partial s$ is the first term of Eq. (1.9)[2]. The other terms of this equation are bounded by assumptions (i) and (ii) of Assumptions 1.1 (see p. 5). Therefore, the first term must also be bounded. Using Eq. (2.39) this term can be represented as

$$\frac{\partial \sigma_{ss}}{\partial s} = \frac{\partial \sigma_h}{\partial s} + \frac{A_s \, (1 - \beta)}{3} s^{-\beta} + o\left(s^{-\beta}\right) \tag{2.40}$$

as $s \to 0$. The second term on the right hand side of this equation approaches infinity as $s \to 0$. In order to cancel this term, it is necessary to assume that

$$\sigma_h = \sigma_0 - \frac{A_s s^{(1-\beta)}}{3} + o\left[s^{(1-\beta)}\right] \tag{2.41}$$

as $s \to 0$. Here σ_0 is independent of s. Equations (2.39) and (2.41) combine to give

$$\sigma_{qq} = \sigma_0 + \frac{(A_q - A_s)}{3} s^{(1-\beta)} + o\left[s^{(1-\beta)}\right], \quad \sigma_{ss} = \sigma_0 + o\left[s^{(1-\beta)}\right],$$
$$\sigma_{\theta\theta} = \sigma_0 + \frac{(A_\theta - A_s)}{3} s^{(1-\beta)} + o\left[s^{(1-\beta)}\right]. \tag{2.42}$$

Substituting this equation and Eq. (2.38) into Eq. (1.9)[1] leads to

$$\left\{ \frac{\partial \sigma_0}{\partial q} + k_0 \left(2\frac{\partial H}{\partial s} + \frac{H}{r}\frac{\partial r}{\partial s} \right) + o\left(1\right) \right\} +$$
$$\left\{ \frac{1}{3} \left[\frac{d(A_q - A_s)}{dq} + \frac{(A_q - A_\theta)}{r}\frac{\partial r}{\partial q} \right] s^{(1-\beta)} + o\left[s^{(1-\beta)}\right] \right\} + \tag{2.43}$$
$$\left\{ 2Hk_1 \left(1 - \beta\right) s^{(1-2\beta)} + o\left[s^{(1-2\beta)}\right] \right\} = 0$$

as $s \to 0$. If $1 - \beta = 1 - 2\beta$ then $\beta = 0$. This contradicts Eq. (2.29). Therefore, it follows from Eq. (2.43) that $1 - 2\beta = 0$ or

$$\beta = \frac{1}{2} \tag{2.44}$$

and

$$\frac{d\left(A_q - A_s\right)}{dq} + \frac{\left(A_q - A_\theta\right)}{r}\frac{\partial r}{\partial q} = 0$$

at $s = 0$. Substituting Eqs. (2.30) and (2.31) into Eq. (1.11) and using Eq. (2.44) it is possible to arrive at Eq. (2.20). The distribution of stresses near the maximum friction surface is found from Eqs. (2.38), (2.42), and (2.44) as

$$\sigma_{qq} = \sigma_0 + \frac{(A_q - A_s)}{3}\sqrt{s} + o\left(\sqrt{s}\right), \quad \sigma_{ss} = \sigma_0 + o\left(\sqrt{s}\right),$$
$$\sigma_{\theta\theta} = \sigma_0 + \frac{(A_\theta - A_s)}{3}\sqrt{s} + o\left(\sqrt{s}\right), \quad \sigma_{qs} = k_0 + k_1 s + o\left(s\right) \tag{2.45}$$

as $s \to 0$. The distribution of the velocity component u_q follows from Eqs. (2.28) and (2.44) in the form

$$u_q = u_0 + u_1\sqrt{s} + o\left(\sqrt{s}\right) \tag{2.46}$$

as $s \to 0$. Using this representation of u_q it is possible to determine the shear strain rate from Eq. (1.10) as

$$\xi_{qs} = \frac{u_1}{4\sqrt{s}} + o\left(\frac{1}{\sqrt{s}}\right) \tag{2.47}$$

as $s \to 0$. Equations (2.32) and (2.44) combine to give

$$\lambda = \lambda_0 s^{-1/2} + o\left(s^{-1/2}\right) \tag{2.48}$$

as $s \to 0$. Substituting this equation along with Eqs. (1.28) and (2.47) into Eq. (1.19) yields

$$\lambda_0 = \frac{u_1}{12k_0}. \tag{2.49}$$

Equations (2.33) and (2.34) combine to give $\xi_{ss} = \lambda_0 A_s + o(1)$ as $s \to 0$. Substituting Eq. (2.49) into this equation and integrating results in

$$u_s = \frac{u_1 A_s}{12k_0}s + o(s) \tag{2.50}$$

as $s \to 0$. It has been taken into account here that u_s should satisfy the boundary condition (1.8).

In contrast to plane strain deformation, the constitutive equations for axisymmetric deformation depend on the yield criterion. The analysis presented in this section is based on the von Mises yield criterion. However, it has been shown in [2] that Eq. (2.20) is valid for quite a general smooth yield criterion and in [1] that this equation is valid for Tresca yield criterion.

2.3   Compression of a Layer Between Rough Plates

As an illustrative example of singular solutions, an approximate solution for compression of a rigid plastic layer between parallel plates is given in this section. This solution can be found in any monograph on plasticity theory, for example [3].

2.3.1 Statement of the Problem

Consider a rigid perfectly plastic layer of thickness $2h$ and width $2w$. The layer is compressed between two parallel plates. The speed of each plate is V. The Cartesian coordinate system (x, y) is chosen as shown in Fig. 2.2. The process has two axes of symmetry, $x = w$ and $y = 0$. Therefore, it is sufficient to consider the domain $0 \le x \le w$ and $0 \le y \le h$.

Let u_x and u_y be the velocity components referred to the Cartesian coordinate system. The exact velocity boundary conditions are

$$u_y = 0 \qquad\qquad (2.51)$$

for $y = 0$ and

$$u_y = -V \qquad\qquad (2.52)$$

for $y = h$. The exact velocity boundary condition at $x = w$ is replaced with the following approximate condition [3]

$$\int_0^h u_x dy = 0. \qquad\qquad (2.53)$$

It is understood here that the velocity component u_x involved in the integrand is calculated at $x = w$.

Let σ_{xx}, σ_{yy} and σ_{xy} be the stress components referred to the Cartesian coordinate system. The exact stress boundary conditions are

$$\sigma_{xy} = 0 \qquad\qquad (2.54)$$

for $y = 0$ and the maximum friction law at $y = h$. The exact stress boundary conditions at $x = 0$ are replaced with the following approximate condition [3]

Fig. 2.2 Compression of a plastic layer between two parallel plates notation

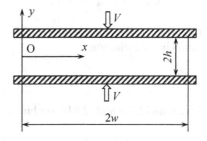

$$\int_0^h \sigma_{xx} dy = 0. \tag{2.55}$$

It is understood here that the stress component σ_{xx} involved in the integrand is calculated at $x = 0$.

In the Cartesian coordinate system, Eqs. (1.1) and (1.2) become

$$\frac{\partial \sigma_{xx}}{\partial x} + \frac{\partial \sigma_{xy}}{\partial y} = 0, \quad \frac{\partial \sigma_{xy}}{\partial x} + \frac{\partial \sigma_{yy}}{\partial y} = 0 \tag{2.56}$$

and

$$\xi_{xx} = \frac{\partial u_x}{\partial x}, \quad \xi_{yy} = \frac{\partial u_y}{\partial y}, \quad \xi_{xy} = \frac{1}{2}\left(\frac{\partial u_x}{\partial y} + \frac{\partial u_y}{\partial x}\right), \tag{2.57}$$

respectively.

It is evident that in the case under consideration the normal distance from the maximum friction surface is

$$s = h - y. \tag{2.58}$$

The coordinate line $s = 0$ (Fig. 1.1) corresponds to $y = h$ and the angle ψ shown in Fig. 1.3 is also the angle which the principal stress direction corresponding to σ_1 makes with the x- axis.

2.3.2   Solution

The yield criterion (1.15) and the associate flow rule (1.16) in the Cartesian coordinate system read

$$\left(\sigma_{xx} - \sigma_{yy}\right)^2 + 4\sigma_{xy}^2 = 4k_0^2. \tag{2.59}$$

and

$$\xi_{xx} = \lambda\left(\sigma_{xx} - \sigma_{yy}\right), \quad \xi_{yy} = \lambda\left(\sigma_{yy} - \sigma_{xx}\right), \quad \xi_{xy} = 2\lambda\sigma_{xy}, \tag{2.60}$$

respectively. By analogy to Eq. (2.1) the yield criterion (2.59) is satisfied by the following substitution

$$\sigma_{xx} = \sigma + k_0 \cos 2\psi, \quad \sigma_{yy} = \sigma - k_0 \cos 2\psi, \quad \sigma_{xy} = k_0 \sin 2\psi. \tag{2.61}$$

Equations (2.60) and (2.61) combine to give

$$\xi_{xx} = 2k_0\lambda \cos 2\psi, \quad \xi_{yy} = -2k_0\lambda \cos 2\psi, \quad \xi_{xy} = 2k_0\lambda \sin 2\psi. \tag{2.62}$$

Eliminating λ between these equations results in

$$\xi_{xx} + \xi_{yy} = 0, \quad \frac{2\xi_{xy}}{\xi_{xx} - \xi_{yy}} = \tan 2\psi \tag{2.63}$$

The direction of flow (Fig. 2.2) dictates that $\sigma_{xy} \geq 0$ at $y = h$ in the range $0 \leq x \leq w$. The maximum friction law (1.29) becomes

$$\psi = \pi/4 \tag{2.64}$$

for $y = h$. Substituting Eq. (2.61) into Eq. (2.56) yields

$$\frac{\partial \sigma}{\partial x} - 2k_0 \sin 2\psi \frac{\partial \psi}{\partial x} + 2k_0 \cos 2\psi \frac{\partial \psi}{\partial y} = 0, \tag{2.65}$$

$$\frac{\partial \sigma}{\partial y} + 2k_0 \sin 2\psi \frac{\partial \psi}{\partial y} + 2k_0 \cos 2\psi \frac{\partial \psi}{\partial x} = 0.$$

In the case of $h/w \ll 1$ it is reasonable to assume that ψ is independent of x [3]. Then, Eq. (2.65) becomes

$$\frac{\partial \sigma}{\partial x} + 2k_0 \cos 2\psi \frac{d\psi}{dy} = 0, \quad \frac{\partial \sigma}{\partial y} + 2k_0 \sin 2\psi \frac{d\psi}{dy} = 0. \tag{2.66}$$

Equation $(2.66)^2$ can be immediately integrated to give $\sigma = k_0 \cos 2\psi + \Phi_1(x)$ where $\Phi_1(x)$ is an arbitrary function of x. Using this solution to eliminate σ in Eq. $(2.66)^1$ and taking into account that the second term of this equation is independent of x it is possible to find that

$$\frac{\sigma}{k_0} = B - Ax + \cos 2\psi \tag{2.67}$$

and

$$2 \cos 2\psi \frac{d\psi}{dy} = A. \tag{2.68}$$

Here A and B are constant. It follows from the boundary condition (2.54) and Eq. (2.61) that $\psi = 0$ at $y = 0$. The solution of Eq. (2.68) satisfying this boundary condition is $\sin 2\psi = Ay$. Using the boundary condition (2.64) it is possible to determine that $A = 1/h$. Then,

$$\sin 2\psi = \frac{y}{h} \tag{2.69}$$

and Eq. (2.67) becomes

$$\frac{\sigma}{k_0} = B - \frac{x}{h} + \cos 2\psi. \tag{2.70}$$

The value of B can be found from the boundary condition (2.55) using Eqs. (2.61), (2.69) and (2.70). However, it is not necessary to determine this value for demonstrating that the solution is singular. Equations (2.57) and (2.63) combine to give

$$\frac{\partial u_x}{\partial x} + \frac{\partial u_y}{\partial y} = 0, \quad \frac{\partial u_x}{\partial y} + \frac{\partial u_y}{\partial x} = \left(\frac{\partial u_x}{\partial x} - \frac{\partial u_y}{\partial y}\right) \tan 2\psi. \qquad (2.71)$$

In the case of $h/w \ll 1$ it is reasonable to assume that u_y is independent of x [3]. Then, Eq. (2.71) becomes

$$\frac{\partial u_x}{\partial x} = -\frac{du_y}{dy}, \quad \frac{\partial u_x}{\partial y} = -2\frac{du_y}{dy} \tan 2\psi. \qquad (2.72)$$

Differentiating Eq. (2.72)^1 with respect to y and Eq. (2.72)^2 with respect to x leads to

$$\frac{\partial^2 u_x}{\partial x \partial y} = -\frac{d^2 u_y}{dy^2}, \quad \frac{\partial^2 u_x}{\partial y \partial x} = 0.$$

Then, using the equation

$$\frac{\partial^2 u_x}{\partial x \partial y} = \frac{\partial^2 u_x}{\partial y \partial x}$$

it is possible to find that

$$\frac{d^2 u_y}{dy^2} = 0.$$

The solution of this equation satisfying the boundary conditions (2.51) and (2.52) is

$$\frac{u_y}{V} = -\frac{y}{h}. \qquad (2.73)$$

Equations (2.72)^2 and (2.73) combine to give

$$\frac{\partial u_x}{\partial y} = \frac{2V}{h} \tan 2\psi. \qquad (2.74)$$

Replacing here differentiation with respect to y with differentiation with respect to ψ by means of Eq. (2.69) yields

$$\frac{\partial u_x}{\partial \psi} = 4V \sin 2\psi. \qquad (2.75)$$

Integrating this equation results in

$$\frac{u_x}{V} = u_0 - 2\cos 2\psi. \qquad (2.76)$$

Here u_0 is an arbitrary function of x. It is seen from Eq. (2.72)[1] that the derivative $\partial u_x / \partial x$ is independent of x. It is therefore evident that u_0 is a linear function of x and Eq. (2.76) becomes

$$\frac{u_x}{V} = a + bx - 2\cos 2\psi \tag{2.77}$$

where a and b are constant. Substituting Eqs. (2.73) and (2.77) into Eq. (2.72)[1] shows that $b = 1/h$. Then, Eq. (2.77) becomes

$$\frac{u_x}{V} = a + \frac{x}{h} - 2\cos 2\psi. \tag{2.78}$$

The value of a can be found from the boundary condition (2.53) using Eqs. (2.69) and (2.78). However, it is not necessary to determine this value for demonstrating that the solution is singular. In particular, it is seen from Eq. (2.73) that $\partial u_y / \partial x = 0$. Therefore, it follows from Eq. (2.57) that $2\xi_{xy} = \partial u_x / \partial y$. Then, using Eq. (2.74)

$$\xi_{xy} = \frac{V}{h} \tan 2\psi. \tag{2.79}$$

Expanding the left hand side of Eq. (2.69) in a series in the vicinity of $\psi = \pi/4$ and using Eq. (2.58) yield

$$s = 2h\left(\psi - \frac{\pi}{4}\right)^2 + o\left[\left(\psi - \frac{\pi}{4}\right)^2\right] \tag{2.80}$$

as $\psi \to \pi/4$. Expanding the right hand side of Eq. (2.79) in a series in the vicinity of $\psi = \pi/4$ and using Eq. (2.80) yield

$$\xi_{xy} = \frac{V}{\sqrt{2hs}} + o\left(\frac{1}{\sqrt{s}}\right) \tag{2.81}$$

as $s \to 0$. Since the strain rate components ξ_{xx} and ξ_{yy} are bounded, Eq. (2.81) coincides with Eq. (2.20).

References

1. Alexandrov S, Richmond O (1998) Asymptotic behavior of the velocity field in the case of axially symmetric flow of a material obeying the Treska condition. Dokl Phys 43:362–364
2. Alexandrov S, Richmond O (2001) Singular plastic flow fields near surfaces of maximum friction stress. Int J Non Linear Mech 36:1–11
3. Hill R (1950) The mathematical theory of plasticity. Clarendon Press, Oxford

Chapter 3
Rigid Viscoplastic Material

3.1 Plane Strain Deformation [6–8]

The (q, s) coordinate system illustrated in Fig. 1.1 will be used. The system of equations consisting of Eqs. (1.1), (1.2), (1.6), and (1.20) is not hyperbolic. The maximum friction law in the form of Eq. (1.31) is adopted. The yield criterion (1.20) is satisfied by the following substitution

$$\sigma_{qq} = \sigma - k \sin 2\gamma, \quad \sigma_{ss} = \sigma + k \sin 2\gamma, \quad \sigma_{qs} = k \cos 2\gamma \tag{3.1}$$

where σ is the stress invariant introduced in Eq. (1.7), γ is the orientation of one of the maximum shear stress direction relative to the q direction and

$$k = k_0 f\left(\xi_{eq}\right). \tag{3.2}$$

The subsequent analysis is restricted to the following class of functions

$$k = k_s - k_\infty \xi_{eq}^{-\delta} + o\left(\xi_{eq}^{-\delta}\right) \tag{3.3}$$

as $\xi_{eq} \to \infty$ where $k_\infty > 0$ and $\delta > 0$. Since $df/d\xi_{eq} \ge 0$ for all ξ_{eq}, it follows from Eqs. (3.1), (3.2) and (3.3) that the boundary condition (1.31) is satisfied if and only if

$$\gamma = 0 \tag{3.4}$$

at $s = 0$. A consequence of Eq. (1.16) is

$$\frac{\xi_{qq} - \xi_{ss}}{\xi_{qs}} = \frac{\sigma_{qq} - \sigma_{ss}}{\sigma_{qs}}. \tag{3.5}$$

© The Author(s) 2018
S. Alexandrov, *Singular Solutions in Plasticity*, SpringerBriefs
in Continuum Mechanics, DOI 10.1007/978-981-10-5227-9_3

Substituting Eq. (3.1) into Eq. (3.5) and using Eqs. (1.2) and (1.6) give

$$\xi_{ss} = -\xi_{qq} = \xi_{qs} \tan 2\gamma. \tag{3.6}$$

It follows from Eqs. (3.4) and (3.1) that $\cos 2\gamma > 0$ in the vicinity of the maximum friction surface. Therefore, substituting Eq. (3.6) into Eq. (1.3) and taking into account that Eqs. (1.27) and (1.16) result in the inequality $\xi_{qs} > 0$ yield

$$\xi_{eq} = \frac{2}{\sqrt{3}} \frac{\xi_{qs}}{\cos 2\gamma}. \tag{3.7}$$

It follows from Eqs. (3.4) and (3.6) that $\xi_{qq} = 0$ at $s = 0$ unless $\xi_{qs} \to \infty$ as $s \to 0$. The condition $\xi_{qq} = 0$ at $s = 0$ is very restrictive. In particular, it is seen from Eqs. (1.2) and (1.8) that this condition results in a constant value of the velocity component u_q over the maximum friction surface. Therefore, if $u_q = 0$ at one point of the friction surface then $u_q = 0$ over the entire surface, i.e. no sliding occurs. Therefore, in what follows it is assumed that $\xi_{qq} \neq 0$ at $s = 0$ and therefore

$$\xi_{qs} \to \infty \tag{3.8}$$

as $s \to 0$. Using assumptions (i) and (iii) of Assumptions 1.1 (see p. 5) along with the boundary condition (1.8) the velocity components are represented as

$$u_q = U_{q0} + U_{q1}s^\beta + o\left(s^\beta\right), \quad u_s = U_{s1}s^\omega + o\left(s^\omega\right) \tag{3.9}$$

as $s \to 0$. Here U_{q0}, U_{q1} and U_{s1} are independent of s, $\beta > 0$ and $\omega > 0$. Substituting Eq. (3.9) into the expression for the shear strain rate in Eq. (1.2) gives

$$2\xi_{qs} = \left[\frac{1}{H} \frac{dU_{s1}}{dq} s^\omega + o\left(s^\omega\right) \right] + \left[U_{q1} \beta s^{\beta-1} + o\left(s^{\beta-1}\right) \right] + O\left(1\right) \tag{3.10}$$

as $s \to 0$. Since $\omega > 0$, the only possibility to satisfy the condition (3.8) is to put

$$1 > \beta > 0. \tag{3.11}$$

It is worthy of note that the inequality $\beta < 0$ contradicts the inequality (1.5). Using Eq. (3.11) it is possible to rewrite Eq. (3.10) as

$$\xi_{qs} = \frac{1}{2} U_{q1} \beta s^{\beta-1} + o\left(s^{\beta-1}\right) \tag{3.12}$$

as $s \to 0$. Substituting Eq. (3.9) into the expression for the normal strain rates in Eq. (1.2) leads to

$$\xi_{qq} = \left[\frac{1}{H} \left(\frac{dU_{q0}}{dq} + \frac{dU_{q1}}{dq} s^\beta \right) + o\left(s^\beta\right) \right] + \left[\frac{U_{s1}}{H} \frac{\partial H}{\partial s} s^\omega + o\left(s^\omega\right) \right], \quad (3.13)$$

$$\xi_{ss} = U_{s1} \omega s^{\omega-1} + o\left(s^{\omega-1}\right)$$

as $s \to 0$. In this equation, H and $\partial H/\partial s$ are understood to be calculated at $s = 0$. Since the strain rate component $\xi_{qq} = O\,(1)$ as $s \to 0$, it follows from Eqs. (1.2) and (1.6) that the strain rate component $\xi_{ss} = O\,(1)$ as $s \to 0$. Then, it is possible to find from Eq. (3.13) that

$$\omega = 1. \quad (3.14)$$

Taking into account assumptions (i) and (iii) of Assumptions 1.1 (see p. 5) as well as the boundary condition (3.4) the angle γ is represented as

$$\gamma = \Phi_1 s^\chi + o\left(s^\chi\right) \quad (3.15)$$

as $s \to 0$. Here Φ_1 is independent of s and $\chi > 0$. Then,

$$\sin 2\gamma = 2\Phi_1 s^\chi + o\left(s^\chi\right), \quad \cos 2\gamma = 1 - 2\Phi_1^2 s^{2\chi} + o\left(s^{2\chi}\right), \quad (3.16)$$

$$\tan 2\gamma = 2\Phi_1 s^\chi + o\left(s^\chi\right)$$

as $s \to 0$. Substituting Eqs. (3.12), (3.13) and (3.16) into Eq. (3.6) and using Eq. (3.14) yield

$$U_{s1} = \Phi_1 U_{q1} \beta s^{\chi+\beta-1} + o\left(s^{\chi+\beta-1}\right) \quad (3.17)$$

as $s \to 0$. This equation is satisfied if and only if

$$\chi + \beta = 1. \quad (3.18)$$

It follows from Eqs. (3.11) and (3.18) that

$$0 < \chi < 1. \quad (3.19)$$

Eliminating ξ_{qs} and $\cos 2\gamma$ in Eq. (3.7) by means of Eqs. (3.12) and (3.16), respectively, gives

$$\xi_{eq} = \frac{\beta U_{q1}}{\sqrt{3}} s^{\beta-1} + o\left(s^{\beta-1}\right) \quad (3.20)$$

as $s \to 0$. Substituting Eq. (3.20) into Eq. (3.3) and using Eq. (3.18) to eliminate β yields

$$k = k_s - k_\infty \left[\frac{\sqrt{3}}{(1-\chi)U_{q1}} \right]^\delta s^{\chi\delta} + o\left(s^{\chi\delta}\right) \quad (3.21)$$

as $s \to 0$.

It is now necessary to consider Eq. (1.1). To this end, the asymptotic representation of the derivatives $\partial\sigma_{qq}/\partial q$, $\partial\sigma_{qs}/\partial q$, $\partial\sigma_{ss}/\partial s$, and $\partial\sigma_{qs}/\partial s$ in the vicinity of the surface $s = 0$ should be found. Differentiating the last equation in Eq. (3.1) with respect to q leads to

$$\frac{\partial\sigma_{qs}}{\partial q} = \frac{\partial k}{\partial q}\cos 2\gamma - 2k\sin 2\gamma\frac{\partial\gamma}{\partial q}. \tag{3.22}$$

Differentiating Eq. (3.21) with respect to q the derivative $\partial k/\partial q$ can be represented as

$$\frac{\partial k}{\partial q} = \frac{k_\infty\delta\,(1-\chi)}{\sqrt{3}}\left[\frac{\sqrt{3}}{(1-\chi)\,U_{q1}}\right]^{(1+\delta)}\frac{dU_{q1}}{dq}s^{\delta\chi} + o\left(s^{\delta\chi}\right) \tag{3.23}$$

as $s \to 0$. Substituting Eqs. (3.3), (3.15), (3.16), and (3.23) into Eq. (3.22) results in

$$\frac{\partial\sigma_{qs}}{\partial q} = \left[\frac{k_\infty\delta}{\sqrt{3}}\left(\frac{\sqrt{3}}{U_{q1}}\right)^{1+\delta}\frac{dU_{q1}}{dq}(1-\chi)^{-\delta}s^{\chi\delta} + o\left(s^{\chi\delta}\right)\right] - \tag{3.24}$$

$$\left[4k_s\Phi_1\frac{d\Phi_1}{dq}s^{2\chi} + o\left(s^{2\chi}\right)\right]$$

as $s \to 0$. Since δ introduced in Eq. (3.3) is positive, it is evident from Eqs. (3.19) and (3.24) that the derivative $\partial\sigma_{qs}/\partial q$ vanishes at $s = 0$. It is seen from Eq. (3.24) that

$$\frac{\partial\sigma_{qs}}{\partial q} = \frac{k_\infty\delta}{\sqrt{3}}\left(\frac{\sqrt{3}}{U_{q1}}\right)^{1+\delta}\frac{dU_{q1}}{dq}(1-\chi)^{-\delta}s^{\chi\delta} + o\left(s^{\chi\delta}\right) \tag{3.25}$$

as $s \to 0$ if $\delta < 2$,

$$\frac{\partial\sigma_{qs}}{\partial q} = -4k_s\Phi_1\frac{d\Phi_1}{dq}s^{2\chi} + o\left(s^{2\chi}\right) \tag{3.26}$$

as $s \to 0$ if $\delta > 2$ and

$$\frac{\partial\sigma_{qs}}{\partial q} = 2\left[\frac{k_\infty}{\sqrt{3}(1-\chi)^2}\left(\frac{\sqrt{3}}{U_{q1}}\right)^3\frac{dU_{q1}}{dq} - 2k_s\Phi_1\frac{d\Phi_1}{dq}\right]s^{2\chi} + o\left(s^{2\chi}\right) \tag{3.27}$$

as $s \to 0$ if $\delta = 2$. Differentiating the first equation in Eq. (3.1) with respect to q gives

$$\frac{\partial\sigma_{qq}}{\partial q} = \frac{\partial\sigma}{\partial q} - \sin 2\gamma\frac{\partial k}{\partial q} - 2k\cos 2\gamma\frac{\partial\gamma}{\partial q}. \tag{3.28}$$

Substituting Eqs. (3.3), (3.15), (3.16), and (3.23) into Eq. (3.28) yields

$$\frac{\partial \sigma_{qq}}{\partial q} = \frac{\partial \sigma}{\partial q} - \left\{ \frac{2k_\infty \delta (1 - \chi) \Phi_1}{\sqrt{3}} \left[\frac{\sqrt{3}}{(1 - \chi) U_{q1}} \right]^{(1+\delta)} \frac{dU_{q1}}{dq} s^{(\delta\chi+\chi)} + o \left[s^{(\delta\chi+\chi)} \right] \right\} -$$
$$\left[2k_s \frac{d\Phi_1}{dq} s^\chi + o \left(s^\chi \right) \right] \tag{3.29}$$

as $s \to 0$. Since both δ and χ are positive, $\delta\chi + \chi > \chi$ and Eq. (3.29) becomes

$$\frac{\partial \sigma_{qq}}{\partial q} = \frac{\partial \sigma}{\partial q} - \left[2k_s \frac{d\Phi_1}{dq} s^\chi + o \left(s^\chi \right) \right] \tag{3.30}$$

as $s \to 0$.

Differentiating the last equation in Eq. (3.1) with respect to s gives

$$\frac{\partial \sigma_{qs}}{\partial s} = \cos 2\gamma \frac{\partial k}{\partial s} - 2k \sin 2\gamma \frac{\partial \gamma}{\partial s}. \tag{3.31}$$

Using Eq. (3.21) the derivative $\partial k / \partial s$ can be represented as

$$\frac{\partial k}{\partial s} = -k_\infty \chi \delta \left[\frac{\sqrt{3}}{(1 - \chi) U_{q1}} \right]^\delta s^{(\chi\delta-1)} + o \left[s^{(\chi\delta-1)} \right] \tag{3.32}$$

as $s \to 0$. Substituting Eqs. (3.3), (3.15), (3.16), and (3.32) into Eq. (3.31) yields

$$\frac{\partial \sigma_{qs}}{\partial s} = - \left\{ k_\infty \chi \delta \left[\frac{\sqrt{3}}{(1 - \chi) U_{q1}} \right]^\delta s^{(\chi\delta-1)} + o \left[s^{(\chi\delta-1)} \right] \right\} -$$
$$\left\{ 4k_s \chi \Phi_1^2 s^{(2\chi-1)} + o \left[s^{(2\chi-1)} \right] \right\}$$

as $s \to 0$. It follows from this equation that

$$\frac{\partial \sigma_{qs}}{\partial s} = -k_\infty \chi \delta \left[\frac{\sqrt{3}}{(1 - \chi) U_{q1}} \right]^\delta s^{(\chi\delta-1)} + o \left[s^{(\chi\delta-1)} \right] \tag{3.33}$$

as $s \to 0$ if $\delta < 2$,

$$\frac{\partial \sigma_{qs}}{\partial s} = -4k_s \Phi_1^2 \chi s^{(2\chi-1)} + o \left[s^{(2\chi-1)} \right] \tag{3.34}$$

as $s \to 0$ if $\delta > 2$ and

$$\frac{\partial \sigma_{qs}}{\partial s} = -2\chi \left[\frac{3k_\infty}{(1-\chi)^2 U_{q1}^2} + 2k_s \Phi_1^2 \right] s^{(2\chi-1)} + o\left[s^{(2\chi-1)} \right] \tag{3.35}$$

as $s \to 0$ if $\delta = 2$.

Differentiating the second equation in Eq. (3.1) with respect to s yields

$$\frac{\partial \sigma_{ss}}{\partial s} = \frac{\partial \sigma}{\partial s} + \sin 2\gamma \frac{\partial k}{\partial s} + 2k \cos 2\gamma \frac{\partial \gamma}{\partial s}. \tag{3.36}$$

Substituting Eqs. (3.3), (3.15), (3.16), and (3.32) into Eq. (3.36) gives

$$\frac{\partial \sigma_{ss}}{\partial s} = \frac{\partial \sigma}{\partial s} - \left\{ 2\Phi_1 k_\infty \chi \delta \left[\frac{\sqrt{3}}{(1-\chi) U_{q1}} \right]^\delta s^{(\chi\delta - 1 + \chi)} + o\left[s^{(\chi\delta - 1 + \chi)} \right] \right\} + $$
$$\left\{ 2k_s \Phi_1 \chi s^{(\chi-1)} + o\left[s^{(\chi-1)} \right] \right\} \tag{3.37}$$

as $s \to 0$. Since the product $\chi\delta$ is positive, $\chi\delta - 1 + \chi > \chi - 1$ and Eq. (3.37) becomes

$$\frac{\partial \sigma_{ss}}{\partial s} = \frac{\partial \sigma}{\partial s} + 2k_s \Phi_1 \chi s^{(\chi-1)} + o\left[s^{(\chi-1)} \right] \tag{3.38}$$

as $s \to 0$.

It is seen from Eqs. (3.25), (3.26), (3.27), (3.33), (3.34), and (3.35) that it is convenient to separately examine the cases $\delta < 2$, $\delta > 2$ and $\delta = 2$. Assume first that $\delta > 2$. In this case, substituting Eqs. (3.1), (3.3), (3.16), (3.26), and (3.38) into the second equation in Eq. (1.1) yields

$$\frac{\partial \sigma}{\partial s} = -2k_s \Phi_1 \chi s^{(\chi-1)} + o\left[s^{(\chi-1)} \right] \tag{3.39}$$

as $s \to 0$. Integrating this equation gives

$$\sigma = \sigma_0 - 2k_s \Phi_1 s^\chi + o\left(s^\chi \right) \tag{3.40}$$

as $s \to 0$. Here σ_0 is an arbitrary function of q. Using Eqs. (3.1), (3.3), (3.30), and (3.34) it is possible to represent the first equation in Eq. (1.1) as

$$\frac{\partial \sigma}{\partial q} + 2k_s \frac{\partial H}{\partial s} - 4Hk_s \Phi_1^2 \chi s^{(2\chi-1)} + o\left[s^{(2\chi-1)} \right] = 0 \tag{3.41}$$

as $s \to 0$. Substituting Eq. (3.40) into Eq. (3.41) gives

$$\frac{d\sigma_0}{dq} + 2k_s \frac{\partial H}{\partial s} - 4Hk_s \Phi_1^2 \chi s^{(2\chi-1)} + o\left[s^{(2\chi-1)} \right] = 0 \tag{3.42}$$

as $s \to 0$. In Eqs. (3.41) and (3.42), it has been taken into account that $2\chi - 1 < \chi$ in the range shown in Eq. (3.19). Moreover, H and $\partial H/\partial s$ are understood to be calculated at $s = 0$. It follows from Eq. (3.42) that

$$\chi = \frac{1}{2}. \tag{3.43}$$

Substituting this equation into Eq. (3.18) yields

$$\beta = \frac{1}{2}. \tag{3.44}$$

Using Eqs. (3.1), (3.16), (3.21), (3.40), and (3.43) the distribution of the stress components σ_{qq}, σ_{ss} and σ_{qs} in the vicinity of maximum friction surfaces is represented as

$$\sigma_{qq} = \sigma_0 - 4k_s \Phi_1 \sqrt{s} + o\left(\sqrt{s}\right), \quad \sigma_{ss} = \sigma_0 + o\left(\sqrt{s}\right), \tag{3.45}$$
$$\sigma_{qs} = k_s - 2k_s \Phi_1^2 s + o(s)$$

as $s \to 0$. The first equation in Eq. (1.1) involves the derivative $\partial \sigma_{qq}/\partial q$. Differentiating the expression for σ_{qq} in Eq. (3.45) with respect to q shows that this derivative contains the term $4k_s (d\Phi_1/dq) \sqrt{s}$. In general, this is the only term of the order $\sqrt{s}$ as $s \to 0$ involved in the first equation in Eq. (1.1). Therefore, it is necessary to assume that Φ_1 is constant. The asymptotic representation of the velocity components u_q and u_s follows from Eqs. (3.9), (3.14) and (3.44) in the form

$$u_q = U_{q0} + U_{q1}\sqrt{s} + o\left(\sqrt{s}\right), \quad u_s = \frac{\Phi_1 U_{q1}}{2}s + o(s) \tag{3.46}$$

as $s \to 0$. Here the coefficient U_{s1} involved in the expression for u_s in Eq. (3.9) has been eliminated by means of Eq. (3.17). The equivalent strain rate is determined from Eqs. (3.20) and (3.44) as

$$\xi_{eq} = \frac{U_{q1}}{2\sqrt{3}\sqrt{s}} + \left(\frac{1}{\sqrt{s}}\right) \tag{3.47}$$

as $s \to 0$.

Examine the case $\delta = 2$. In this case, substituting Eqs. (3.1), (3.16), (3.21), (3.27), and (3.38) into the second equation in Eq. (1.1) yields

$$\frac{\partial \sigma}{\partial s} = -2k_s \Phi_1 \chi s^{(\chi-1)} + o\left[s^{(\chi-1)}\right] \tag{3.48}$$

as $s \to 0$. It has been taken into account here that $\chi - 1 < 2\chi$ in the range shown in Eq. (3.19). Integrating Eq. (3.48) gives

$$\sigma = \sigma_0 - 2k_s \Phi_1 s^\chi + o\left(s^\chi\right) \tag{3.49}$$

as $s \to 0$. As before, σ_0 is an arbitrary function of q. Using Eqs. (3.1), (3.16), (3.21), (3.23), and (3.35) it is possible to represent the first equation in Eq. (1.1) as

$$\frac{\partial \sigma}{\partial q} + 2k_s \frac{\partial H}{\partial s} - 2H\chi \left[\frac{3k_\infty}{(1-\chi)^2 U_{q1}^2} + 2k_s \Phi_1^2 \right] s^{(2\chi-1)} + o\left[s^{(2\chi-1)} \right] \quad (3.50)$$

as $s \to 0$. In this equation, H and $\partial H/\partial s$ are understood to be calculated at $s = 0$. Substituting Eq. (3.49) into Eq. (3.50) yields

$$\frac{d\sigma_0}{dq} + 2k_s \frac{\partial H}{\partial s} - 2H\chi \left[\frac{3k_\infty}{(1-\chi)^2 U_{q1}^2} + 2k_s \Phi_1^2 \right] s^{(2\chi-1)} + o\left[s^{(2\chi-1)} \right]$$

as $s \to 0$. It follows from this equation that

$$\chi = \frac{1}{2}. \quad (3.51)$$

Substituting this equation into Eq. (3.18) yields

$$\beta = \frac{1}{2}. \quad (3.52)$$

Using Eqs. (3.1), (3.16), (3.21), (3.49) and (3.51) the distribution of the stress components σ_{qq}, σ_{ss} and σ_{qs} in the vicinity of maximum friction surfaces is represented as

$$\sigma_{qq} = \sigma_0 - 4k_s \Phi_1 \sqrt{s} + o\left(\sqrt{s} \right), \quad \sigma_{ss} = \sigma_0 + o\left(\sqrt{s} \right), \quad (3.53)$$

$$\sigma_{qs} = k_s - \left[\frac{3k_\infty}{(1-\chi)^2 U_{q1}^2} + 2k_s \Phi_1^2 \right] s + o(s)$$

as $s \to 0$. As in the case $\delta > 2$, it is possible to show that Φ_1 is constant. The asymptotic representation of the velocity components u_q and u_s follows from Eqs. (3.9), (3.14) and (3.52) in the form

$$u_q = U_{q0} + U_{q1}\sqrt{s} + o\left(\sqrt{s} \right), \quad u_s = \frac{\Phi_1 U_{q1}}{2} s + o(s) \quad (3.54)$$

as $s \to 0$. Here the coefficient U_{s1} involved in the expression for u_s in Eq. (3.9) has been eliminated by means of Eq. (3.17). The equivalent strain rate is determined from Eqs. (3.20) and (3.52) as

$$\xi_{eq} = \frac{U_{q1}}{2\sqrt{3}\sqrt{s}} + \left(\frac{1}{\sqrt{s}} \right) \quad (3.55)$$

as $s \to 0$.

Examine the case $\delta < 2$. In this case, substituting Eqs. (3.1), (3.16), (3.21), (3.25), and (3.38) into the second equation in Eq. (1.1) yields

$$\frac{\partial \sigma}{\partial s} + 2k_s \Phi_1 \chi s^{(\chi-1)} + o\left[s^{(\chi-1)}\right] = 0 \tag{3.56}$$

as $s \to 0$. It has been taken into account here that $\chi - 1 < \delta \chi$ because δ is positive and χ belongs to the range shown in Eq. (3.19). Integrating Eq. (3.56) gives

$$\sigma = \sigma_0 - 2k_s \Phi_1 s^\chi + o\left(s^\chi\right) \tag{3.57}$$

as $s \to 0$. As before, σ_0 is an arbitrary function of q. Using Eqs. (3.1), (3.16), (3.21), (3.30), and (3.33) it is possible to represent the first equation in Eq. (1.1) as

$$\frac{\partial \sigma}{\partial q} + 2k_s \left\{ \frac{\partial H}{\partial s} - \frac{H k_\infty \delta \chi}{(1-\chi)^\delta} \left(\frac{\sqrt{3}}{U_{q1}}\right)^\delta \left\{ s^{(\delta\chi-1)} + o\left[s^{(\delta\chi-1)}\right] \right\} \right\} = 0 \tag{3.58}$$

as $s \to 0$. It has been taken into account here that $\delta \chi - 1 < \chi$ because δ is positive and χ belongs to the range shown in Eq. (3.19). Also, H and $\partial H/\partial s$ are understood to be calculated at $s = 0$. Substituting Eq. (3.57) into Eq. (3.58) yields

$$\frac{d\sigma_0}{dq} + 2k_s \left\{ \frac{\partial H}{\partial s} - \frac{H k_\infty \delta \chi}{(1-\chi)^\delta} \left(\frac{\sqrt{3}}{U_{q1}}\right)^\delta \left\{ s^{(\delta\chi-1)} + o\left[s^{(\delta\chi-1)}\right] \right\} \right\} = 0 \tag{3.59}$$

as $s \to 0$. It follows from this equation that

$$\chi = \frac{1}{\delta}. \tag{3.60}$$

Substituting Eq. (3.60) into Eq. (3.18) results in

$$\beta = 1 - \frac{1}{\delta}. \tag{3.61}$$

It is seen from Eqs. (3.19) and (3.60) that the present analysis is valid if $\delta > 1$. If $0 < \delta \le 1$ then the regime of sliding at maximum friction surfaces is impossible. Such qualitative behaviour of viscoplastic solutions for models with a saturation stress is similar to that of viscoplastic solutions for models with no saturation stress [3, 5].

Using Eqs. (3.1), (3.16), (3.21), (3.57) and (3.60) the distribution of the stress components σ_{qq}, σ_{ss} and σ_{qs} in the vicinity of maximum friction surfaces is represented as

$$\sigma_{qq} = \sigma_0 - 4k_s\Phi_1 s^{1/\delta} + o\left(s^{1/\delta}\right), \quad \sigma_{ss} = \sigma_0 + o\left(s^{1/\delta}\right), \qquad (3.62)$$

$$\sigma_{qs} = k_s - k_\infty \left[\frac{\sqrt{3\delta}}{(\delta-1)\,U_{q1}}\right]^\delta s + o\left(s\right)$$

as $s \to 0$. The first equation in Eq. (1.1) involves the derivative $\partial\sigma_{qq}/\partial q$. Differentiating the expression for σ_{qq} in Eq. (3.62) with respect to q shows that this derivative contains the term $4k_s\left(d\Phi_1/dq\right)s^{1/\delta}$. This is the only term of the order $s^{1/\delta}$ as $s \to 0$ involved in the first equation in Eq. (1.1). Therefore, it is necessary to assume that Φ_1 is constant. The asymptotic representation of the velocity components u_q and u_s follows from Eqs. (3.9), (3.14) and (3.61) in the form

$$u_q = U_{q0} + U_{q1}s^{(1-1/\delta)} + o\left[s^{(1-1/\delta)}\right], \quad u_s = \Phi_1 U_{q1}\left(1 - \frac{1}{\delta}\right)s + o\left(s\right) \quad (3.63)$$

as $s \to 0$. Here the coefficient U_{s1} involved in the expression for u_s in Eq. (3.9) has been eliminated by means of Eq. (3.17). The equivalent strain rate is determined from Eqs. (3.20) and (3.61) as

$$\xi_{eq} = \frac{U_{q1}\left(\delta - 1\right)}{\sqrt{3\delta}}s^{-1/\delta} + o\left(s^{-1/\delta}\right) \qquad (3.64)$$

as $s \to 0$.

3.2 Axisymmeric Deformation [9]

Axisymmetric deformation is described by Eqs. (1.9), (1.10), (1.19), and (1.21). It is assumed that Eqs. (3.2) and (3.3) are valid. The maximum friction law in the form of Eq. (1.31) is adopted. The (q, θ, s) coordinate system illustrated in Fig. 1.2 will be used. Equations (1.14), (1.21), (1.23) and (1.31) combine to give

$$\sigma_h = \sigma_{\alpha\alpha} = \sigma_{ss} = \sigma_{\theta\theta} = \sigma_0 \qquad (3.65)$$

at $s = 0$. Here σ_0 is an arbitrary function of q. Substituting Eq. (3.65) into Eq. (1.19) for $\xi_{\theta\theta}$ shows that $\xi_{\theta\theta} = 0$ unless

$$\lambda \to \infty \qquad (3.66)$$

as $s \to 0$. The condition $\xi_{\theta\theta} = 0$ contradicts Eq. (1.10) for $\xi_{\theta\theta}$ if, at a generic point M of the friction surface, the regime of sliding occurs (the velocity vector is not equal to zero) and the tangent to the friction surface is not parallel to the z- axis, which is the axis of symmetry (Fig. 2.1). Therefore, Eq. (3.66) is in general valid. Then, it follows from Eqs. (1.19) and (1.31) that

$$\xi_{qs} \to \infty \qquad (3.67)$$

as $s \to 0$. Using assumption (ii) of Assumptions 1.1 (see p. 5) it is possible to conclude that the derivative $\partial u_s / \partial q$ is bounded at $s = 0$. Then, it follows from Eqs. (1.10) and (3.67) that

$$\frac{\partial u_q}{\partial s} \to \infty \qquad (3.68)$$

as $s \to 0$. Using assumption (iii) of Assumptions 1.1 (see p. 5) it is possible to represent u_q as

$$u_q = u_0 + u_1 s^\beta + o\left(s^\beta\right) \qquad (3.69)$$

as $s \to 0$. Here u_0 and u_1 are independent of s. It follows from assumption (i) of Assumptions 1.1 (see p. 5) that $\beta > 0$. On the other hand, it is seen from Eqs. (3.68) and (3.69) that $\beta < 1$. Hence,

$$0 < \beta < 1. \qquad (3.70)$$

Using assumptions (i) and (ii) of Assumptions 1.1 (see p. 5) it is possible to find from Eq. (1.10) that the strain rate components ξ_{qq} and $\xi_{\theta\theta}$ are bounded as $s \to 0$. Then, it follows from Eqs. (1.10) and (1.13) that the strain rate component ξ_{ss} is bounded as $s \to 0$. Therefore,

$$\xi_{qq} = O(1), \ \xi_{ss} = O(1) \text{ and } \xi_{\theta\theta} = O(1) \qquad (3.71)$$

as $s \to 0$. Substituting Eq. (3.69) into Eq. (1.10) and then into Eq. (1.11) lead to

$$\xi_{qs} = O\left(s^{\beta-1}\right) \text{ and } \xi_{eq} = O\left(s^{\beta-1}\right) \qquad (3.72)$$

as $s \to 0$. Since stresses are bounded, it is seen from Eqs. (1.12), (3.70), (3.71) and (3.72) that the inequality (1.5) is satisfied. Equations (1.19) for ξ_{qs}, (1.31) and (3.72) combine to give $\lambda = O\left(s^{\beta-1}\right)$ as $s \to 0$. This equation can be rewritten as

$$\lambda = \lambda_0 s^{\beta-1} + o\left(s^{\beta-1}\right) \qquad (3.73)$$

as $s \to 0$. Here λ_0 is independent of s. Substituting Eq. (3.73) into Eq. (1.19) leads to

$$\begin{aligned}
\xi_{qq} &= \left[\lambda_0 s^{\beta-1} + o\left(s^{\beta-1}\right)\right]\left(2\sigma_{qq} - \sigma_{ss} - \sigma_{\theta\theta}\right), \\
\xi_{ss} &= \left[\lambda_0 s^{\beta-1} + o\left(s^{\beta-1}\right)\right]\left(2\sigma_{ss} - \sigma_{\theta\theta} - \sigma_{qq}\right), \\
\xi_{\theta\theta} &= \left[\lambda_0 s^{\beta-1} + o\left(s^{\beta-1}\right)\right]\left(2\sigma_{\theta\theta} - \sigma_{ss} - \sigma_{qq}\right).
\end{aligned} \qquad (3.74)$$

as $s \to 0$. Then, it follows from Eq. (3.71) that

$$2\sigma_{qq} - \sigma_{ss} - \sigma_{\theta\theta} = A_q s^{1-\beta} + o\left(s^{1-\beta}\right),$$
$$2\sigma_{ss} - \sigma_{\theta\theta} - \sigma_{qq} = A_s s^{1-\beta} + o\left(s^{1-\beta}\right), \qquad (3.75)$$
$$2\sigma_{\theta\theta} - \sigma_{qq} - \sigma_{ss} = A_\theta s^{1-\beta} + o\left(s^{1-\beta}\right)$$

as $s \to 0$. Here A_q, A_s and A_θ are independent of s. Equation (3.75) can be transformed to

$$\sigma_{qq} - \sigma_{ss} = \frac{\left(A_q - A_s\right)}{3} s^{1-\beta} + o\left(s^{1-\beta}\right),$$
$$\sigma_{ss} - \sigma_{\theta\theta} = \frac{\left(A_s - A_\theta\right)}{3} s^{1-\beta} + o\left(s^{1-\beta}\right), \qquad (3.76)$$
$$\sigma_{\theta\theta} - \sigma_{qq} = \frac{\left(A_\theta - A_q\right)}{3} s^{1-\beta} + o\left(s^{1-\beta}\right)$$

as $s \to 0$. Using assumption (iii) of Assumptions 1.1 (see p. 5) and taking into account Eq. (1.31) it is possible to represent σ_{qs} as

$$\sigma_{\alpha s} = k_s + A_{qs} s^\omega + o\left(s^\omega\right) \qquad (3.77)$$

as $s \to 0$. Here $\omega > 0$ and A_{qs} is independent of s. Using Eqs. (3.74), (3.75), and (3.77) the strain rate components are represented as

$$\xi_{qq} = \lambda_0 A_q + o\left(1\right), \quad \xi_{ss} = \lambda_0 A_s + o\left(1\right), \qquad (3.78)$$
$$\xi_{\theta\theta} = \lambda_0 A_\theta + o\left(1\right), \quad \xi_{qs} = 3\lambda_0 k_s s^{(\beta-1)} + o\left[s^{(\beta-1)}\right]$$

as $s \to 0$. Equations (1.11) and (3.78) combine to give

$$\xi_{eq} = 2\sqrt{3}\lambda_0 k_s s^{(\beta-1)} + o\left[s^{(\beta-1)}\right] \qquad (3.79)$$

as $s \to 0$. Substituting Eq. (3.79) into Eq. (3.3) leads to

$$k = k_s - k_\infty \left(2\sqrt{3}\lambda_0 k_s\right)^{-\delta} s^{\delta(1-\beta)} + o\left[s^{\delta(1-\beta)}\right] \qquad (3.80)$$

as $s \rightarrow 0$. Substituting Eqs. (3.76), (3.77) and (3.80) into Eq. (1.21) and using Eqs. (3.3) and (3.79) result in

$$\left\{ \left[\frac{\left(A_q - A_s\right)^2 + \left(A_s - A_\theta\right)^2 + \left(A_\theta - A_q\right)^2}{9} \right] s^{2(1-\beta)} + o\left[s^{2(1-\beta)}\right] \right\} + \quad (3.81)$$

$$\left[12k_s A_{qs} s^\omega + o\left(s^\omega\right)\right] = -12k_s k_\infty \left(2\sqrt{3}\lambda_0 k_s\right)^{-\delta} s^{\delta(1-\beta)} + o\left[s^{\delta(1-\beta)}\right]$$

as $s \rightarrow 0$. It is worthy of note that the coefficient of $s^{2(1-\beta)}$ never vanishes. Using Eq. (1.14) it is possible to rewrite Eq. (3.75) as

$$\sigma_{qq} - \sigma_h = \frac{A_q}{3} s^{1-\beta} + o\left(s^{1-\beta}\right), \quad \sigma_{ss} - \sigma_h = \frac{A_s}{3} s^{1-\beta} + o\left(s^{1-\beta}\right), \quad (3.82)$$

$$\sigma_{\theta\theta} - \sigma_h = \frac{A_\theta}{3} s^{1-\beta} + o\left(s^{1-\beta}\right)$$

as $s \rightarrow 0$.

It follows from Eq. (3.81) that it is necessary to examine the following cases

$$\delta < 2 \qquad (3.83)$$

$$\delta > 2, \qquad (3.84)$$

$$\delta = 2, \qquad (3.85)$$

Consider the case $\delta > 2$. In this case Eq. (3.81) gives

$$\omega = 2(1 - \beta), \quad A_{qs} = -\frac{\left[\left(A_q - A_s\right)^2 + \left(A_s - A_\theta\right)^2 + \left(A_\theta - A_q\right)^2\right]}{108k_s}. \qquad (3.86)$$

It is worthy of note that $A_{qs} \neq 0$. Equation (3.77) becomes

$$\sigma_{qs} = k_s + A_{qs} s^{2(1-\beta)} + o\left[s^{2(1-\beta)}\right] \qquad (3.87)$$

as $s \rightarrow 0$. Substituting Eqs. (3.82) and (3.87) into the second equation in Eq. (1.9) and using Eq. (3.76) results in

$$\frac{\partial \sigma_h}{\partial s} + \frac{A_s (1 - \beta)}{3} s^{-\beta} + o\left(s^{-\beta}\right) = 0$$

as $s \to 0$. It has been taken into account here that the value of β should be in the range shown in Eq. (3.70). Integrating this equation gives

$$\sigma_h = \sigma_0 - \frac{A_s}{3} s^{(1-\beta)} + o\left[s^{(1-\beta)}\right] \tag{3.88}$$

as $s \to 0$. Here Eq. (3.65) has been taken into account. Substituting Eqs. (3.82) and (3.87) into the first equation in Eq. (1.9) and using Eq. (3.76) yields

$$\frac{\partial \sigma_h}{\partial q} + k_s \left(2\frac{\partial H}{\partial s} + \frac{H}{r}\frac{\partial r}{\partial s}\right) + \left\{2HA_{qs}(1-\beta) s^{(1-2\beta)} + o\left[s^{(1-2\beta)}\right]\right\} \tag{3.89}$$

as $s \to 0$. It has been taken into account here that the value of β should be in the range shown in Eq. (3.70). Also, H, $\partial H/\partial s$, r, and $\partial r/\partial s$ are understood to be calculated at $s = 0$. Substituting Eq. (3.88) into Eq. (3.89) leads to

$$\frac{d\sigma_0}{dq} + k_s \left(2\frac{\partial H}{\partial s} + \frac{H}{r}\frac{\partial r}{\partial s}\right) + \left\{2HA_{qs}(1-\beta) s^{(1-2\beta)} + o\left[s^{(1-2\beta)}\right]\right\} \tag{3.90}$$

as $s \to 0$. It follows from this equation that

$$\beta = \frac{1}{2}. \tag{3.91}$$

Substituting this equation into Eq. (3.88) supplies the distribution of the hydrostatic stress in the vicinity of maximum friction surfaces as

$$\sigma_h = \sigma_0 - \frac{A_s}{3}\sqrt{s} + o\left(\sqrt{s}\right) \tag{3.92}$$

as $s \to 0$. The distribution of the stress components is determined from Eqs. (3.82), (3.87), (3.91) and (3.92) in the form

$$\sigma_{qs} = k_s + A_{qs}s + o(s), \quad \sigma_{qq} = \sigma_0 + \frac{(A_q - A_s)}{3}\sqrt{s} + o\left(\sqrt{s}\right), \tag{3.93}$$

$$\sigma_{ss} = \sigma_0 + o\left(\sqrt{s}\right), \quad \sigma_{\theta\theta} = \sigma_0 + \frac{(A_\theta - A_s)}{3}\sqrt{s} + o\left(\sqrt{s}\right)$$

as $s \to 0$. Here A_{qs} can be eliminated by means of Eq. (3.86). Moreover, the coefficients A_q, A_θ and A_s satisfy the relation

$$A_q + A_\theta + A_s = 0 \tag{3.94}$$

as follows from Eq. (3.75). Equation (3.94) can be used for eliminating one of these coefficients in Eq. (3.93). Moreover, substituting Eq. (3.93) into the first equation in Eq. (1.9) shows that this equation contains the term

$$\left[\frac{d\left(A_q - A_s\right)}{dq} + \frac{\left(A_q - A_\theta\right)}{r} \frac{\partial r}{\partial q} \right] \frac{\sqrt{s}}{3}.$$

This is the only term of the order $\sqrt{s}$ as $s \to 0$ involved in the first equation in Eq. (1.9). Therefore, it is necessary to assume that

$$\frac{d\left(A_q - A_s\right)}{dq} + \frac{\left(A_q - A_\theta\right)}{r} \frac{\partial r}{\partial q} = 0. \qquad (3.95)$$

In this equation, r and $\partial r/\partial s$ are understood to be calculated at $s = 0$. Using Eq. (3.91) the velocity component u_q given in Eq. (3.69) is represented as

$$u_q = u_0 + u_1 \sqrt{s} + o\left(\sqrt{s}\right). \qquad (3.96)$$

as $s \to 0$. Substituting Eqs. (3.78), (3.91) and (3.96) into Eq. (1.10) for ξ_{qs} leads to

$$u_1 = 12\lambda_0 k_s. \qquad (3.97)$$

Equations (3.96) and (3.97) combine to give

$$u_q = u_0 + 12\lambda_0 k_s \sqrt{s} + o\left(\sqrt{s}\right) \qquad (3.98)$$

as $s \to 0$. In order to find the asymptotic representation of the velocity component u_s in the vicinity of maximum friction surfaces, it is necessary to eliminate ξ_{ss} in Eq. (1.10) by means of Eq. (3.78). As a result,

$$\frac{\partial u_s}{\partial s} = \lambda_0 A_s + o\left(1\right) \qquad (3.99)$$

as $s \to 0$. Integrating this equation and using the boundary condition (1.8) yields

$$u_s = \lambda_0 A_s s + o\left(s\right) \qquad (3.100)$$

as $s \to 0$. Finally, the equivalent strain rate is represented by means of Eqs. (3.79) and (3.91) as

$$\xi_{eq} = \frac{2\sqrt{3}\lambda_0 k_s}{\sqrt{s}} + o\left(\frac{1}{\sqrt{s}}\right) \qquad (3.101)$$

as $s \to 0$.

Consider the case $\delta = 2$. In this case Eq. (3.81) gives

$$\omega = 2(1 - \beta), \tag{3.102}$$

$$A_{qs} = -\frac{k_\infty}{12\lambda_0^2 k_s^2} - \frac{\left[(A_q - A_s)^2 + (A_s - A_\theta)^2 + (A_\theta - A_q)^2\right]}{108 k_s}.$$

It is worthy of note that the coefficient A_{qs} never vanishes. Equation (3.77) becomes

$$\sigma_{qs} = k_s + A_{qs} s^{2(1-\beta)} + o\left[s^{2(1-\beta)}\right] \tag{3.103}$$

as $s \to 0$. Substituting Eqs. (3.82) and (3.103) into the second equation in Eq. (1.9) and using Eq. (3.76) results in

$$\frac{\partial \sigma_h}{\partial s} + \frac{A_s (1 - \beta)}{3} s^{-\beta} + o\left(s^{-\beta}\right) = 0$$

as $s \to 0$. It has been taken into account here that the value of β should be in the range shown in Eq. (3.70). Integrating this equation gives

$$\sigma_h = \sigma_0 - \frac{A_s}{3} s^{(1-\beta)} + o\left[s^{(1-\beta)}\right] \tag{3.104}$$

as $s \to 0$. Here Eq. (3.65) has been taken into account. Substituting Eqs. (3.82) and (3.103) into the first equation in Eq. (1.9) and using Eq. (3.76) yields

$$\frac{\partial \sigma_h}{\partial q} + k_s \left(2\frac{\partial H}{\partial s} + \frac{H}{r}\frac{\partial r}{\partial s}\right) + \left\{2H A_{qs} (1 - \beta) s^{(1-2\beta)} + o\left[s^{(1-2\beta)}\right]\right\} \tag{3.105}$$

as $s \to 0$. It has been taken into account here that the value of β should be in the range shown in Eq. (3.70). Also, H, $\partial H / \partial s$, r, and $\partial r / \partial s$ are understood to be calculated at $s = 0$. Substituting Eq. (3.104) into Eq. (3.105) leads to

$$\frac{d\sigma_0}{dq} + k_s \left(2\frac{\partial H}{\partial s} + \frac{H}{r}\frac{\partial r}{\partial s}\right) + \left\{2H A_{qs} (1 - \beta) s^{(1-2\beta)} + o\left[s^{(1-2\beta)}\right]\right\} \tag{3.106}$$

as $s \to 0$. It follows from this equation that

$$\beta = \frac{1}{2}. \tag{3.107}$$

Substituting this equation into Eq. (3.88) supplies the distribution of the hydrostatic stress in the vicinity of maximum friction surfaces as

$$\sigma_h = \sigma_0 - \frac{A_s}{3}\sqrt{s} + o\left(\sqrt{s}\right) \tag{3.108}$$

as $s \to 0$. The distribution of the stress components is determined from Eqs. (3.82), (3.103) and (3.108) in the form

$$\sigma_{qs} = k_s + A_{qs}s + o(s), \quad \sigma_{qq} = \sigma_0 + \frac{(A_q - A_s)}{s}\sqrt{s} + o(\sqrt{s}), \qquad (3.109)$$

$$\sigma_{ss} = \sigma_0 + o(\sqrt{s}), \quad \sigma_{\theta\theta} = \sigma_0 + \frac{(A_\theta - A_s)}{s}\sqrt{s} + o(\sqrt{s})$$

as $s \to 0$. Here A_{qs} can be eliminated by means of Eq. (3.102). Moreover, the coefficients A_q, A_θ and A_s satisfy Eq. (3.94). This equation can be used for eliminating one of these coefficients in Eq. (3.109). As in the case $\delta > 2$, it is possible to show that the coefficients A_q, A_θ and A_s should satisfy Eq. (3.95). Using Eq. (3.91) the velocity component u_q given in Eq. (3.69) is represented as

$$u_q = u_0 + u_1\sqrt{s} + o(\sqrt{s}) \qquad (3.110)$$

as $s \to 0$. Substituting Eqs. (3.78) and (3.110) into Eq. (1.10) for ξ_{qs} leads to Eq. (3.97). Then, Eqs. (3.97) and (3.110) combine to give

$$u_q = u_0 + 12\lambda_0 k_s\sqrt{s} + o(\sqrt{s}) \qquad (3.111)$$

as $s \to 0$. In order to find the asymptotic representation of the velocity component u_s in the vicinity of maximum friction surfaces, it is necessary to eliminate ξ_{ss} in Eq. (1.10) by means of Eq. (3.78). As a result,

$$\frac{\partial u_s}{\partial s} = \lambda_0 A_s + o(1) \qquad (3.112)$$

as $s \to 0$. Integrating this equation and using the boundary condition (1.8) yields

$$u_s = \lambda_0 A_s s + o(s) \qquad (3.113)$$

as $s \to 0$. Finally, the equivalent strain rate is represented by means of Eqs. (3.79) and (3.107) as

$$\xi_{eq} = \frac{2\sqrt{3}\lambda_0 k_s}{\sqrt{s}} + o\left(\frac{1}{\sqrt{s}}\right) \qquad (3.114)$$

as $s \to 0$.

Consider the case $\delta < 2$. In this case Eq. (3.81) gives

$$\omega = \delta(1 - \beta), \quad A_{qs} = -k_\infty\left(2\sqrt{3}\lambda_0 k_s\right)^{-\delta}. \qquad (3.115)$$

Equation (3.77) becomes

$$\sigma_{qs} = k_s + A_{qs} s^{\delta(1-\beta)} + o\left[s^{\delta(1-\beta)}\right] \tag{3.116}$$

as $s \to 0$. Substituting Eq. (3.82) and Eq. (3.116) into the second equation in Eq. (1.9) and using Eq. (3.76) results in

$$\frac{\partial \sigma_h}{\partial s} + \frac{A_s(1-\beta)}{3} s^{-\beta} + o\left(s^{-\beta}\right) = 0 \tag{3.117}$$

as $s \to 0$. It has been taken into account here that δ is positive and the value of β should be in the range shown in Eq. (3.70). Integrating Eq. (3.117) and using Eq. (3.65) gives

$$\sigma_h = \sigma_0 - \frac{A_s}{3} s^{(1-\beta)} + o\left[s^{(1-\beta)}\right] \tag{3.118}$$

as $s \to 0$. Substituting Eqs. (3.82) and (3.116) into the first equation in Eq. (1.9) and using Eq. (3.76) results in

$$\frac{\partial \sigma_h}{\partial q} + k_s \left(2\frac{\partial H}{\partial s} + \frac{H}{r}\frac{\partial r}{\partial s}\right) + H A_{qs} \delta (1-\beta) s^{(\delta-\delta\beta-1)} + o\left[s^{(\delta-\delta\beta-1)}\right] \tag{3.119}$$

as $s \to 0$. It has been taken into account here that δ is positive and the value of β should be in the range shown in Eq. (3.70). Therefore, $1 - \beta > \delta - \delta\beta - 1$. Also, H, $\partial H/\partial s$, r, and $\partial r/\partial s$ are understood to be calculated at $s = 0$. Substituting Eq. (3.118) into Eq. (3.119) gives

$$\frac{d\sigma_0}{dq} + k_s \left(2\frac{\partial H}{\partial s} + \frac{H}{r}\frac{\partial r}{\partial s}\right) + H A_{qs} \delta (1-\beta) s^{(\delta-\delta\beta-1)} + o\left[s^{(\delta-\delta\beta-1)}\right] \tag{3.120}$$

as $s \to 0$. It follows from this equation that

$$\beta = 1 - \frac{1}{\delta}. \tag{3.121}$$

Equations (3.115) and (3.121) combine to give

$$\omega = 1. \tag{3.122}$$

Substituting this equation into Eq. (3.118) supplies the distribution of the hydrostatic stress in the vicinity of maximum friction surfaces as

$$\sigma_h = \sigma_0 - \frac{A_s}{3} s^{1/\delta} + o\left(s^{1/\delta}\right) \tag{3.123}$$

as $s \to 0$. The distribution of the stress components is determined from Eqs. (3.82), (3.116) and (3.123) in the form

$$\sigma_{qs} = k_s + A_{qs}s + o\,(s)\,, \quad \sigma_{qq} = \sigma_0 + \frac{(A_q - A_s)}{3}s^{1/\delta} + o\left(s^{1/\delta}\right), \qquad (3.124)$$

$$\sigma_{ss} = \sigma_0 + o\left(s^{1/\delta}\right), \quad \sigma_{\theta\theta} = \sigma_0 + \frac{(A_\theta - A_s)}{3}s^{1/\delta} + o\left(s^{1/\delta}\right)$$

as $s \to 0$. Here A_{qs} can be eliminated by means of Eq. (3.115). Moreover, the coefficients A_q, A_θ and A_s satisfy Eq. (3.94). This equation can be used for eliminating one of these coefficients in Eq. (3.124). Moreover, substituting Eq. (3.124) into the first equation in Eq.(1.9) shows that this equation contains the term

$$\frac{1}{3}\left[\frac{d\,(A_q - A_s)}{dq} + \frac{(A_q - A_\theta)}{r}\frac{\partial r}{\partial q}\right]s^{1/\delta}$$

This is the only term of the order $s^{1/\delta}$ as $s \to 0$ involved in the first equation in Eq. (1.9). Therefore, it is necessary to assume that the coefficients A_q, A_θ and A_s satisfy Eq. (3.95). Using Eq. (3.121) the velocity component u_q given in Eq. (3.69) is represented as

$$u_q = u_0 + u_1 s^{(1-1/\delta)} + o\left[s^{(1-1/\delta)}\right] \qquad (3.125)$$

as $s \to 0$. Substituting Eqs. (3.78), (3.121) and (3.125) into Eq. (1.10) for ξ_{qs} leads to

$$u_1 = \frac{6\lambda_0 k_s \delta}{\delta - 1}.$$

Then, this equation and (3.125) combine to give

$$u_q = u_0 + \frac{6\lambda_0 k_s \delta}{(\delta - 1)}s^{(1-1/\delta)} + o\left[s^{(1-1/\delta)}\right] \qquad (3.126)$$

as $s \to 0$. In order to find the asymptotic representation of the velocity component u_s in the vicinity of maximum friction surfaces, it is necessary to eliminate ξ_{ss} in Eq. (1.10) by means of Eq. (3.78). As a result,

$$\frac{\partial u_s}{\partial s} = \lambda_0 A_s + o\,(1) \qquad (3.127)$$

as $s \to 0$. Integrating this equation and using the boundary condition (1.8) yields

$$u_s = \lambda_0 A_s s + o\,(s) \qquad (3.128)$$

as $s \to 0$. Finally, the equivalent strain rate is represented by means of Eqs. (3.79) and (3.121) as

$$\xi_{eq} = 2\sqrt{3}\lambda_0 k_s s^{-1/\delta} + o\left(s^{-1/\delta}\right) \tag{3.129}$$

as $s \to 0$.

3.3 Compression of a Layer Between Rough Plates

The statement of the problem is provided in Sect. 2.3.1. Below is given the solution of this problem assuming that

$$\frac{k}{k_s} = \frac{\left(\xi_{eq}/\xi_0\right)^{\delta} + t}{\left(\xi_{eq}/\xi_0\right)^{\delta} + 1} \tag{3.130}$$

where $t = k_0/k_s$ and ξ_0 is a material constant. It is assumed that

$$\delta \geq 2 \tag{3.131}$$

The shear yield stress k is expressed in terms of the function $f\left(\xi_{eq}\right)$ involved in Eq. (1.20) as $k = k_0 f\left(\xi_{eq}\right)$. Comparing Eqs. (3.3) and (3.130) shows that $k_\infty = (1-t)\xi_0^{\delta}$. Equation (3.130) has been proposed in [10] using results of numerous independent experiments. The solution of the same boundary value problem for the dependence

$$k = k_s \left[1 - (1-t)\exp\left(-\frac{\xi_{eq}}{\xi_0}\right)\right] \tag{3.132}$$

has been found in [4]. Equation (3.132) satisfies Eq. (1.23) but does not satisfy Eq. (3.3). Nevertheless, it has been found in [4] that the asymptotic representation of the solution derived in Sect. 3.1 for the case $\delta > 2$ is valid. A solution for a model satisfying Eq. (1.22) has been provided in [1]. This solution does not exist if the maximum friction law is adopted. An axisymmetric analogue of the problem formulated in Sect. 2.3.1 for the dependence given by Eq. (3.130) has been solved in [2].

The yield criterion (1.20) and the associate flow rule (1.16) in the Cartesian coordinate system read

$$\left(\sigma_{xx} - \sigma_{yy}\right)^2 + 4\sigma_{xy}^2 = 4k^2. \tag{3.133}$$

and

$$\xi_{xx} = \lambda\left(\sigma_{xx} - \sigma_{yy}\right), \quad \xi_{yy} = \lambda\left(\sigma_{yy} - \sigma_{xx}\right), \quad \xi_{xy} = 2\lambda\sigma_{xy}, \tag{3.134}$$

respectively. By analogy to Eq. (2.61) the yield criterion (3.133) is satisfied by the following substitution

$$\sigma_{xx} = \sigma + k \cos 2\psi, \quad \sigma_{yy} = \sigma - k \cos 2\psi, \quad \sigma_{xy} = k \sin 2\psi. \tag{3.135}$$

Equations (3.134) and (3.135) combine to give

$$\xi_{xx} = 2k\lambda \cos 2\psi, \quad \xi_{yy} = -2k\lambda \cos 2\psi, \quad \xi_{xy} = 2k\lambda \sin 2\psi. \tag{3.136}$$

Eliminating λ between these equations results in

$$\xi_{xx} + \xi_{yy} = 0, \quad \frac{2\xi_{xy}}{\xi_{xx} - \xi_{yy}} = \tan 2\psi. \tag{3.137}$$

Using this equation the equivalent strain rate defined in Eq. (1.3) can be expressed as

$$\xi_{eq} = \frac{2}{\sqrt{3}} \left| \frac{\xi_{yy}}{\cos 2\psi} \right|. \tag{3.138}$$

The direction of flow (Fig. 2.2) dictates that $\sigma_{xy} \geq 0$ at $y = h$ in the range $0 \leq x \leq w$. In particular, it follows from Eq. (3.135) that a consequence of the maximum friction law (1.31) is

$$\psi = \pi/4 \tag{3.139}$$

for $y = h$. It is evident that $\xi_{yy} < 0$. Therefore, taking into account Eqs. (3.135) and (3.136) the boundary condition (2.54) becomes

$$\psi = 0 \tag{3.140}$$

for $y = 0$. Equation (3.138) can be rewritten as

$$\xi_{eq} = -\frac{2}{\sqrt{3}} \left(\frac{\xi_{yy}}{\cos 2\psi} \right). \tag{3.141}$$

By analogy to the solution for rigid perfectly plastic material (Sect. 2.3.2) it is assumed that

$$\frac{u_y}{V} = -\frac{y}{h}. \tag{3.142}$$

Then, the boundary conditions (2.51) and (2.52) are satisfied. It follows from Eqs. (2.57) and (3.142) that $\xi_{yy} = V/h$ and, then, from the first equation in Eq. (3.137) that $\xi_{xx} = V/h$. Using this equation to eliminate ξ_{xx} in Eq. (2.57) and integrating gives

$$\frac{u_x}{V} = \frac{x}{h} + V_0(y) \tag{3.143}$$

where $V_0(y)$ is an arbitrary function of y. Substituting Eqs. (3.142) and (3.143) into Eq. (3.137)^2 gives

$$\frac{dV_0}{dy} = \frac{2}{h} \tan 2\psi. \tag{3.144}$$

Substituting Eq. (3.142) into Eq. (3.141) and the resulting relation into Eq. (3.130) yield

$$\frac{k}{k_s} = \frac{g + t\cos^\delta 2\psi}{g + \cos^\delta 2\psi}, \quad g = \left(\frac{2V}{\sqrt{3}h\xi_0}\right)^\delta. \tag{3.145}$$

It is seen from Eq. (3.144) that ψ is independent of x and, then, from Eq. (3.145) that k is independent of x. In this case, substituting Eq. (3.135) into Eq. (2.56) and using Eq. (3.145) give

$$\frac{\partial \sigma}{\partial x} + \frac{d (k \sin 2\psi)}{dy} = 0, \quad \frac{\partial (\sigma - k \cos 2\psi)}{\partial y} = 0. \tag{3.146}$$

The solution of Eq. (3.146)^2 is

$$\sigma - k \cos 2\psi = -k_s A (x) \tag{3.147}$$

where $A (x)$ is an arbitrary function of x. Substituting Eq. (3.147) into Eq. (3.146)^1 leads to

$$k_s \frac{dA}{dx} = \frac{d (k \sin 2\psi)}{dy}. \tag{3.148}$$

The left hand side of this equation is a function only x and its right hand side is a function only y. Therefore, it follows from Eq. (3.148) that

$$k_s \frac{dA}{dx} = \frac{d (k \sin 2\psi)}{dy} = B \tag{3.149}$$

where B is constant. Integrating Eq. (3.149) and using the boundary condition (3.140) and Eq. (3.145) give

$$k_s A = Bx + k_s B_0, \quad \sin 2\psi \left(\frac{g + t\cos^\delta 2\psi}{g + \cos^\delta 2\psi}\right) = \frac{B}{k_s} y. \tag{3.150}$$

Here B_0 is constant. It follows from Eqs. (3.139) and (3.150)^2 that $B = k_s / h$. Then, Eq. (3.150) becomes

$$A = \frac{x}{h} + B_0, \quad \sin 2\psi \left(\frac{g + t\cos^\delta 2\psi}{g + \cos^\delta 2\psi}\right) = \frac{y}{h}. \tag{3.151}$$

The second equation provides the dependence of ψ on y. Therefore, this equation and Eq. (3.144) combine to give the dependence of dV_0/dy on y. Expanding the left hand side of Eq. (3.151) in a series in the vicinity of $\psi = \pi/4$ results in

$$2\left(\psi - \frac{\pi}{4}\right)^2 + \frac{2^\delta (1-t)}{g}\left(\psi - \frac{\pi}{4}\right)^\delta + o\left[\left(\psi - \frac{\pi}{4}\right)^2\right] = 1 - \frac{y}{h} \qquad (3.152)$$

as $\psi \to \pi/4$. It is worthy of note that the representation of the last term on the left hand side is independent of the value of δ due to the assumption in Eq. (3.131). However, the coefficient of the leading term on the left hand side of Eq. (3.152) depends on the value of δ. In particular, this coefficient is equal to two if $\delta > 2$ and to $2 + 4(1 - t)/g$ if $\delta = 2$. In either case

$$1 - \frac{y}{h} = O\left[\left(\psi - \frac{\pi}{4}\right)^2\right] \qquad (3.153)$$

as $\psi \to \pi/4$. Therefore, Eq. (3.141) in which ξ_{yy} is eliminated by means of Eqs. (2.57) and (3.142) becomes

$$\xi_{eq} = O\left(\frac{1}{\sqrt{h-y}}\right) \qquad (3.154)$$

as $y \to h$. This equation coincides with Eqs. (3.47) and (3.55) since $s = h - y$ in the case under consideration.

References

1. Adams MJ, Briscoe MJ, Corfield GM, Lawrence CJ, Papathanasiou TD (1997) An analysis of the plane-strain compression of viscous materials. Trans ASME J Appl Mech 64:420–424
2. Aleksandrov SE, Lyamina EA, Tuan NM (2016) Compression of an axisymmetric layer on a rigid mandrel in creep. Mech Solids 51:188–196
3. Alexandrov S, Alexandrova N (2000) On the maximum friction law in viscoplasticity. Mech Time Depend Mater 4:99–104
4. Alexandrov SE, Baranova ID, Mishuris G (2008) Compression of a viscoelastic layer between rough parallel plates. Mech Solids 43:863–869
5. Alexandrov S, Danilov V, Chikanova N (2000) On the stagnation zone in a simulation of axisymmetric pressure metal forming under creep. Mech Solids 35:127–129
6. Alexandrov S, Mishuris G (2007) Viscoplasticity with a saturation stress: distinguished features of the model. Arch Appl Mech 77:35–47
7. Alexandrov S, Mishuris G (2009) Qualitative behaviour of viscoplastic solutions in the vicinity of maximum-friction surfaces. J Eng Math 65:143–156
8. Alexandrov S, Mustafa Y (2013) Singular solutions in viscoplasticity under plane strain conditions. Meccanica 48:2203–2208
9. Alexandrov S, Mustafa Y (2015) Quasi-static axially symmetric viscoplastic flows near very rough walls. Appl Math Model 39:4599–4606
10. Shesterikov SA, Yumasheva MA (1984) More precise specification of the equation of state in creep theory. Mech Solids 19(1):86–91

Chapter 4
Anisotropic Rigid Plastic Material

4.1 Plane Strain Deformation [1]

The (q, s) coordinate system illustrated in Fig. 1.1 will be used. The system of equations consisting of Eqs. (1.1), (1.2), (1.25), and (1.26) is hyperbolic [5]. It follows from Eqs. (1.7) and (1.24) that

$$\sigma_{qq} = \sigma + t \quad \text{and} \quad \sigma_{ss} = \sigma - t. \tag{4.1}$$

Eliminating σ_{qq}, σ_{ss} and σ_{qs} in Eq. (1.1) by means of Eqs. (1.24) and (4.1) leads to

$$\frac{\partial \sigma}{\partial q} + \frac{\partial t}{\partial q} + H\frac{\partial \tau}{\partial s} + 2\tau\frac{\partial H}{\partial s} = 0, \quad H\frac{\partial \sigma}{\partial s} - H\frac{\partial t}{\partial s} + \frac{\partial \tau}{\partial q} - 2t\frac{\partial H}{\partial s} = 0. \tag{4.2}$$

It is seen from the geometry of Fig. 1.4 that

$$dt = -dl\cos 2\gamma, \quad d\tau = -dl\sin 2\gamma. \tag{4.3}$$

Equations (4.2) and (4.3) combine to give

$$\frac{\partial \sigma}{\partial q} - \frac{\partial l}{\partial q}\cos 2\gamma - H\frac{\partial l}{\partial s}\sin 2\gamma + 2\tau\frac{\partial H}{\partial s} = 0, \tag{4.4}$$

$$H\frac{\partial \sigma}{\partial s} + H\frac{\partial l}{\partial s}\cos 2\gamma - \frac{\partial l}{\partial q}\sin 2\gamma - 2t\frac{\partial H}{\partial s} = 0.$$

The maximum friction law is taken in the form of Eq. (1.32). Consequently, $\gamma = 0$ in Fig. 1.4 and the yield contour is orthogonal to the t- axis at points of the maximum friction surface. Consider a class of yield surfaces satisfying the condition $\gamma = 0$ at $t = 0$ (Fig. 4.1). Then,

$$t = 0 \quad \text{and} \quad \tau = \tau_0 \tag{4.5}$$

© The Author(s) 2018
S. Alexandrov, *Singular Solutions in Plasticity*, SpringerBriefs
in Continuum Mechanics, DOI 10.1007/978-981-10-5227-9_4

Fig. 4.1 Yield contour in
Mohr plane when the surface
$s = 0$ coincides with a
maximum friction surface

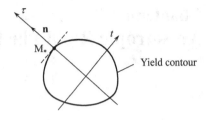

Yield contour

at points of the friction surface. Here τ_0 is the shear yield stress relative to the
coordinate system chosen, a material constant. Therefore, Eq. (1.25) in the vicinity
of point M_* can be rewritten as $\tau = \tau_0 + \Phi(t)$ where $d\Phi/dt = 0$ at $t = 0$. In what
follows, it is assumed that the function $\Phi(t)$ can be represented by a Taylor series in
the vicinity of point M_*. It is obvious that this class of functions is large enough for
all possible engineering applications, unless a piece-wise smooth yield contour is of
interest. Under the assumption made, the yield criterion given in Eq. (1.25) becomes

$$\tau = \tau_0 + \sum_{n=m}^{\infty} a_n t^n. \tag{4.6}$$

where $m \geq 2$. The function $\Phi(t) = \sum_{n=m}^{\infty} a_n t^n$ must be concave in the vicinity of
point M_* (Fig. 4.1). A necessary and sufficient condition for this on some interval
$[-t_0, t_0]$ is $d^2\Phi/dt^2 \leq 0$ for all t in $[-t_0, t_0]$. Therefore, m must be an even integer.
If $\xi_{qs} < \infty$ then $\xi_{qq} - \xi_{ss} = 0$ at $\gamma = 0$, as follows from $(1.26)^2$. Then, this equation
and $(1.26)^1$ result in $\xi_{qq} = \xi_{ss} = 0$. Those are the characteristic relations [5] and,
therefore, the friction surface is not an envelope of characteristics. Thus it is necessary
to assume that $\xi_{qs} \to \infty$ as $s \to 0$. Using assumption (iii) of Assumptions 1.1 (see
p. 5) it is possible to represent ξ_{qs} and γ as

$$\xi_{qs} = \xi_0 s^{-\omega} + o\left(s^{-\omega}\right),$$
$$\gamma = \gamma_0 s^{\chi} + o\left(s^{\chi}\right) \tag{4.7}$$

as $s \to 0$. Here ξ_0 and γ_0 are independent of s. Also

$$\omega > 0 \quad \text{and} \quad \chi > 0. \tag{4.8}$$

Since the shear stress in the (q, s) coordinate system does not vanish at points of
the maximum friction surface, it follows from Eqs. (1.4) and (4.7) that the inequality
(1.5) is satisfied if $\omega < 1$. Therefore, using Eq. (4.8)

$$1 > \omega > 0. \tag{4.9}$$

Assumption (ii) of Assumptions 1.1 (see p. 5) and Eq. (1.2) for ξ_{qq} show that this strain rate component is bounded in the vicinity of point M. Here point M is a point of the friction surface corresponding to point M_* shown in Fig. 4.1. Then, it follows from Eq. (1.26)[1] that the strain rate component ξ_{ss} is bounded as well. Therefore, substituting Eq. (4.7) into (1.26)[2] yields $s^{\chi-\omega} = O(1)$ as $s \to 0$. Then, $\chi = \omega$ and Eq. (4.7)[2] becomes

$$\gamma = \gamma_0 s^\omega + o(s^\omega) \tag{4.10}$$

as $s \to 0$. Then,

$$\cos 2\gamma = 1 - 2\gamma_0^2 s^{2\omega} + o\left(s^{2\omega}\right), \quad \sin 2\gamma = 2\gamma_0 s^\omega + o\left(s^\omega\right), \quad \tan 2\gamma = 2\gamma_0 s^\omega + o\left(s^\omega\right) \tag{4.11}$$

as $s \to 0$. Using assumption (iii) of Assumptions 1.1 (see p. 5) and Eq. (4.5) it is possible to represent σ, t, τ, and l as

$$\begin{aligned} \sigma &= \sigma_0 + \sigma_1 s^\beta + o\left(s^\beta\right), \quad t = t_1 s^\mu + o\left(s^\mu\right), \\ \tau &= \tau_0 + \tau_1 s^\lambda + o\left(s^\lambda\right), \quad l = l_0 + l_1 s^\alpha + o\left(s^\alpha\right) \end{aligned} \tag{4.12}$$

as $s \to 0$. Here $\sigma_0, \sigma_1, t_1, \tau_1, l_0$, and l_1 are independent of s. Also, $\beta > 0, \mu > 0, \lambda > 0$, and $\alpha > 0$. Substituting Eq. (4.12) into Eq. (4.6) gives $\tau_1 s^\lambda + o\left(s^\lambda\right) = a_m t_1^m s^{m\mu} + o(s^{m\mu})$ as $s \to 0$. It follows from this equation that

$$\lambda = m\mu \quad \text{and} \quad \tau_1 = a_m t_1^m. \tag{4.13}$$

Substituting Eqs. (4.11) and (4.12) into Eq. (4.3) yields

$$\begin{aligned} \mu &= \alpha, \quad \lambda = \alpha + \omega, \\ \frac{dl_0}{dq} &= 0, \quad t_1 = -l_1, \quad \frac{d\tau_1}{dq} = -2\gamma_0 \frac{dl_1}{dq}, \quad \tau_1 (\alpha + \omega) = -2\alpha l_1 \gamma_0. \end{aligned} \tag{4.14}$$

Then, Eq. (4.12) can be rewritten as

$$\begin{aligned} \sigma &= \Sigma_0 + \Sigma_1 s^\beta + o\left(s^\beta\right), \quad t = t_1 s^\alpha + o\left(s^\alpha\right), \\ \tau &= \tau_0 + \tau_1 s^{(\alpha+\omega)} + o\left[s^{(\alpha+\omega)}\right], \quad l = l_0 + l_1 s^\alpha + o(s^\alpha) \end{aligned} \tag{4.15}$$

as $s \to 0$. Substituting Eqs. (4.11), (4.14) and (4.15) into Eq. (4.4) gives

$$\begin{aligned} &\left[\frac{d\Sigma_0}{dq} + \frac{d\Sigma_1}{dq} s^\beta + \left(s^\beta\right)\right] - \left[\frac{dl_1}{dq} s^\alpha + \left(s^\alpha\right)\right] - \\ &H\left\{2l_1 \alpha \gamma_0 s^{(\omega+\alpha-1)} + o\left[s^{(\omega+\alpha-1)}\right]\right\} + 2\tau_0 \frac{\partial H}{\partial s} = 0, \\ &\left\{H\Sigma_1 \beta s^{(\beta-1)} + o\left[s^{(\beta-1)}\right]\right\} + \left\{Hl_1 \alpha s^{(\alpha-1)} + o\left[s^{(\alpha-1)}\right]\right\} - \\ &\left\{2\gamma_0 \frac{dl_1}{dq} s^{(\alpha+\omega)} + o\left[s^{(\alpha+\omega)}\right]\right\} = 0 \end{aligned} \tag{4.16}$$

as $s \to 0$. In this equation, H and $\partial H / \partial s$ are understood to be calculated at $s = 0$. Assume that

$$\alpha \neq \beta. \tag{4.17}$$

Then,

$$\alpha \geq 1 \quad \text{and} \quad \beta \geq 1. \tag{4.18}$$

It is seen from the second equation in Eq. (4.16) that $\alpha + \omega = \alpha - 1$ or $\alpha + \omega = \beta - 1$. The former contradicts Eq. (4.9). Therefore, it is necessary to put

$$\beta = 1 + \alpha + \omega. \tag{4.19}$$

It follows from the first equation in Eq. (4.16) that

$$\alpha + \omega - 1 = \alpha \tag{4.20}$$

or

$$\alpha + \omega - 1 = 0 \tag{4.21}$$

or

$$\alpha + \omega - 1 = \beta. \tag{4.22}$$

Equation (4.20) contradicts Eq. (4.9). Equation (4.21) contradicts Eq. (4.9) due to Eq. (4.18). Equations (4.19) and (4.22) are not compatible. Therefore, the assumption in Eq. (4.17) is not true and

$$\alpha = \beta. \tag{4.23}$$

In this case Eq. (4.21) does not lead to any contradiction with Eq. (4.9). Therefore, it follows from Eqs. (4.13) and (4.14) that

$$\alpha = \frac{1}{m}. \tag{4.24}$$

Then, Eqs. (4.13), (4.14), (4.21), (4.23) and (4.24) combine to give

$$\beta = \frac{1}{m}, \quad \mu = \frac{1}{m}, \quad \lambda = 1, \quad \omega = 1 - \frac{1}{m}. \tag{4.25}$$

Moreover, taking into account Eqs. (4.24) and (4.25) it is possible to find from the second equation in Eq. (4.16) and, then, from Eq. (4.14) that

$$\Sigma_1 = -l_1 \quad \text{and} \quad \Sigma_1 = t_1. \tag{4.26}$$

Substituting Eqs. (4.13), (4.24), (4.25) and (4.26) into Eq. (4.12) and the resulting expressions into Eqs. (1.7) and (1.24) yield

$$\sigma_{qq} = \Sigma_0 + 2\Sigma_1 s^{1/m} + o\left(s^{1/m}\right), \quad \sigma_{ss} = \Sigma_0 + o\left(s^{1/m}\right),$$
$$\sigma_{qs} = \tau_0 + a_m \Sigma_1^m s + o\left(s\right) \tag{4.27}$$

as $s \to 0$. In order to find the asymptotic representation of the velocities in the vicinity of the maximum friction surface, it is necessary to consider Eq. (1.26). In particular, taking into account the first equation in Eq. (1.26) it is possible to rewrite the second equation in Eq. (1.26) as $2\xi_{ss} = \xi_{qs} \tan 2\gamma$. Eliminating in this equation ξ_{qs} and $\tan 2\gamma$ by means of Eqs. (4.7) and (4.11) results in $\xi_{ss} = \xi_0 \gamma_0 + o\left(1\right)$ as $s \to 0$. Eliminating in this equation ξ_{ss} by means of (1.2) and integrating gives

$$u_s = \xi_0 \gamma_0 s + o\left(s\right) \tag{4.28}$$

as $s \to 0$. It has been taken into account here that this velocity component should satisfy the boundary condition (1.8). It follows from Eqs. (1.2) and (4.7) that

$$\frac{\partial u_q}{\partial s} = 2\xi_0 s^{-\omega} + o\left(s^{-\omega}\right) \tag{4.29}$$

as $s \to 0$. Integrating and using Eq. (4.25) for eliminating ω yields

$$u_q = U_{q0} + 2m\xi_0 s^{1/m} + o\left(s^{1/m}\right) \tag{4.30}$$

as $s \to 0$. Here U_{q0} is independent of s. Since the normal strain rates in the (q, s) coordinate system are bounded as $s \to 0$, it follows from Eqs. (1.3), (4.7) and (4.25) that

$$\xi_{eq} = \frac{2\xi_0}{\sqrt{3}} s^{-(1-1/m)} + o\left[s^{-(1-1/m)}\right] \tag{4.31}$$

as $s \to 0$.

In terms of t and τ the isotropic yield criterion considered in Sect. 2.1 is written as $t^2 + \tau^2 = \tau_0^2$. Therefore, $\tau = \tau_0 - t^2/(2\tau_0) + o\left(t^2\right)$ as $t \to 0$. Comparing this expansion and Eq. (4.6) shows that $m = 2$. Then, it follows from Eq. (4.25) that $\omega = 1/2$. This result coincides with that obtained in Sect. 2.1. The quadratic yield criterion for orthotropic materials proposed in [4] can be written in the form $At^2 + \tau^2 = \tau_0^2$ where $A > 0$. It is evident that $\omega = 1/2$ in this case as well. Any other yield criterion of the class considered can be treated in a similar matter. In order to determine the exact asymptotic behavior of solutions, it is just necessary to expand the corresponding function in a series in the vicinity of point M_* (Fig. 4.1) and to find the value of m using Eq. (4.6). Then, the value of ω is given by Eq. (4.25).

4.2   Compression of a Layer Between Rough Plates [2]

This boundary value problem has been formulated in Sect. 2.3.1. A number of solutions to this boundary value problem for several material models have been given in [3]. One of the solutions presented in this paper is for instantaneous compression of inhomogeneous anisotropic strips. In order to illustrate the general theory presented in Sect. 4.1, it is sufficient to consider homogeneous strips. Moreover, the solution [3] is valid for quite a general anisotropic yield criterion. However, the present section is restricted to the quadratic yield criterion proposed in [4]. The main result of the present section has been obtained in [2]. In [3], the (x, y) coordinate system has been chosen as shown in Fig. 4.2. Therefore, this coordinate system is used in the present section.

The plane strain yield criterion in terms of the stresses $\sigma_{\zeta\zeta}$, $\sigma_{\eta\eta}$ and $\sigma_{\zeta\eta}$ referred to the principal axes of anisotropy (ζ, η) is

$$\left(\sigma_{\zeta\zeta} - \sigma_{\eta\eta}\right)^2 + 4(1 - c)\sigma_{\zeta\eta}^2 = 4(1 - c)T^2 \tag{4.32}$$

where T is the shear yield stress in the $\zeta\eta$-plane and c is defined as

$$c = 1 - \frac{F + G}{4T^2(FG + GH + HF)}.$$

The parameters involved in this equation are expressible in terms of the yield stresses in respect of the principal axes of anisotropy as

$$2F = \frac{1}{Y^2} + \frac{1}{Z^2} - \frac{1}{X^2}, \ 2G = \frac{1}{Z^2} + \frac{1}{X^2} - \frac{1}{Y^2}, \ 2H = \frac{1}{X^2} + \frac{1}{Y^2} - \frac{1}{Z^2}$$

where X, Y, and Z are the tensile yield stresses in the $\zeta-, \eta-$ and thickness directions, respectively. Theoretically, the value of c can vary in the interval $-\infty < c < 1$. The isotropic material is obtained at $c = 0$. It has been shown in the previous section

Fig. 4.2 Compression of a plastic layer between two parallel plates—notation

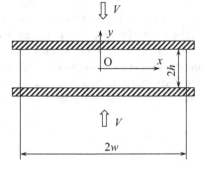

that $m = 2$ in Eq. (4.25) if the yield criterion (4.32) is adopted. Equation (4.32) is satisfied by the following substitution

$$\sigma_{\zeta\zeta} = \sigma - T\sqrt{1-c}\sin 2\varphi, \quad \sigma_{\eta\eta} = \sigma + T\sqrt{1-c}\sin 2\varphi,$$
$$\sigma_{\zeta\eta} = T\cos 2\varphi \tag{4.33}$$

where σ and φ are new unknowns. In particular, σ is the stress invariant introduced in Eq. (1.7). Let θ be the anti-clockwise angular rotation of the x- axis from the ζ- axis. For the purpose of the present study it is sufficient to consider the case $\theta = $ constant. The transformation equations for stress components and Eq. (4.33) give

$$\sigma_{xx} = \sigma - T\sqrt{1-c}\sin 2\varphi\cos 2\theta + T\cos 2\varphi\sin 2\theta,$$
$$\sigma_{yy} = \sigma + T\sqrt{1-c}\sin 2\varphi\cos 2\theta - T\cos 2\varphi\sin 2\theta, \tag{4.34}$$
$$\sigma_{xy} = T\sqrt{1-c}\sin 2\varphi\sin 2\theta + T\cos 2\varphi\cos 2\theta.$$

The maximum friction law is adopted at the surfaces $y = \pm h$ (Fig. 4.2). Therefore, the q- line shown in Fig. 1.3 is straight and parallel to the x- axis. Thus γ is the angle between the x- axis and a characteristic direction. Equation (1.32) is valid at one of the friction surfaces and the condition $|\gamma| = \pi/2$ at the other. In either case,

$$|\cot 2\gamma| \rightarrow \infty \tag{4.35}$$

as $y \rightarrow \pm h$. It has been shown in [5] that the slopes γ and $\gamma + \pi/2$ of the characteristics can be found from the equation $\cot 2\gamma = \partial s_{xy}/\partial \sigma_{xy}$ where $2s_{xy} = \sigma_{xx} - \sigma_{yy}$. Using Eq. (4.34) it is possible to find that

$$\cot 2\gamma = \frac{\sqrt{1-c}\cos 2\varphi\cos 2\theta + \sin 2\varphi\sin 2\theta}{-\sqrt{1-c}\cos 2\varphi\sin 2\theta + \sin 2\varphi\cos 2\theta}. \tag{4.36}$$

Equations (4.35) and (4.36) combine to give

$$\sqrt{1-c}\tan 2\theta = \tan 2\varphi \tag{4.37}$$

at $y = \pm h$. Because of symmetry, it is sufficient to consider the range $0 \le \theta \le \pi/4$. Therefore,

$$\sin 2\theta \ge 0, \quad \cos 2\theta \ge 0 \tag{4.38}$$

and, with the use of Eq. (4.37),

$$\tan 2\varphi \ge 0. \tag{4.39}$$

The direction of flow dictates that

$$\sigma_{xy} < 0 \tag{4.40}$$

at $y = h$ and

$$\sigma_{xy} > 0 \qquad (4.41)$$

at $y = -h$.

Consider the maximum friction surface $y = h$. Assume that $\varphi = \varphi_u$ at $y = h$. It follows from Eqs. (4.34) and (4.40) that

$$\sqrt{1 - c} \sin 2\varphi_u \sin 2\theta + \cos 2\varphi_u \cos 2\theta < 0. \qquad (4.42)$$

Using Eq. (4.38) and eliminating $\tan 2\theta$ in Eq. (4.42) by means of Eq. (4.37) result in the inequality

$$\cos 2\varphi_u < 0. \qquad (4.43)$$

Combining this inequality, Eqs. (4.34) and (4.37) yields the following stress boundary condition at $y = h$

$$\sigma_{xy} = -T \cos 2\theta \sqrt{1 + (1 - c) \tan^2 2\theta}. \qquad (4.44)$$

This condition is equivalent to the maximum friction law.

Consider the maximum friction surface $y = -h$. Assume that $\varphi = \varphi_l$ at $y = -h$. It follows from Eqs. (4.34) and (4.41) that

$$\sqrt{1 - c} \sin 2\varphi_l \sin 2\theta + \cos 2\varphi_l \cos 2\theta > 0. \qquad (4.45)$$

Using Eq. (4.38) and eliminating $\tan 2\theta$ in Eq. (4.45) by means of Eq. (4.37) result in the inequality

$$\cos 2\varphi_l > 0. \qquad (4.46)$$

Combining this inequality, Eqs. (4.34) and (4.37) yields the following stress boundary condition at $y = -h$

$$\sigma_{xy} = T \cos 2\theta \sqrt{1 + (1 - c) \tan^2 2\theta}. \qquad (4.47)$$

This condition is equivalent to the maximum friction law.

The solution given in [3] is valid for the boundary conditions given in Eqs. (4.44) and (4.47). It follows from this solution that

$$\sigma_{xy} = -T \cos 2\theta \sqrt{1 + (1 - c) \tan^2 2\theta} \frac{y}{h},$$
$$\xi_{xx} = -\xi_{yy} = \frac{V}{h}, \quad \xi_{xy} = -\frac{V}{h} \cot 2\gamma. \qquad (4.48)$$

The equivalent strain rate defined in Eq. (1.3) is found with the use of Eqs. (4.36) and (4.48) as

$$\xi_{eq} = \sqrt{\frac{2}{3}}\sqrt{\xi_{xx}^2 + \xi_{yy}^2 + 2\xi_{xy}^2} =$$
$$\sqrt{\frac{2}{3}}\frac{V}{h}\frac{\sqrt{2 - c(1 + \cos 4\varphi)}}{\left|-\sqrt{1 - c}\cos 2\varphi \sin 2\theta + \sin 2\varphi \cos 2\theta\right|}. \tag{4.49}$$

Since Eq. (4.37) is satisfied at $\varphi = \varphi_u$ and $\varphi = \varphi_l$, it is evident from Eq. (4.49) that $\xi_{eq} \to \infty$ as $y \to h$ and $y \to -h$. Moreover, expanding the denominator in Eq. (4.49) in series and using Eq. (4.37) yield

$$-\sqrt{1 - c}\cos 2\varphi \sin 2\theta + \sin 2\varphi \cos 2\theta =$$
$$-2\cos 2\theta \sqrt{(1 - c)\tan^2 2\theta + 1}(\varphi - \varphi_u) + o(\varphi - \varphi_u) \tag{4.50}$$

as $\varphi \to \varphi_u$ and

$$-\sqrt{1 - c}\cos 2\varphi \sin 2\theta + \sin 2\varphi \cos 2\theta =$$
$$2\cos 2\theta \sqrt{(1 - c)\tan^2 2\theta + 1}(\varphi - \varphi_l) + o(\varphi - \varphi_l) \tag{4.51}$$

as $\varphi \to \varphi_l$. Combining Eqs. (4.49), (4.50) and (4.51) gives with the use of Eq. (4.37)

$$\xi_{eq} = \frac{V}{\sqrt{3}h}\frac{\sqrt{1 - c}}{\sqrt{1 - c\sin^2 2\theta}\,|\varphi - \varphi_u|} + o\left[(\varphi - \varphi_u)^{-1}\right] \tag{4.52}$$

as $\varphi \to \varphi_u$ and

$$\xi_{eq} = \frac{V}{\sqrt{3}h}\frac{\sqrt{1 - c}}{\sqrt{1 - c\sin^2 2\theta}\,|\varphi - \varphi_l|} + o\left[(\varphi - \varphi_l)^{-1}\right] \tag{4.53}$$

as $\varphi \to \varphi_l$. It follows from Eqs. (4.34) and (4.48) that

$$\frac{dy}{d\varphi} = -\frac{2h\left(\sqrt{1 - c}\tan 2\theta \cos 2\varphi - \sin 2\varphi\right)}{\sqrt{1 + (1 - c)\tan^2 2\theta}}. \tag{4.54}$$

It is evident from Eqs. (4.37) and (4.54) that $dy/d\varphi = 0$ at $\varphi = \varphi_u$ and $\varphi = \varphi_l$. Expanding the right hand side of Eq. (4.54) in series and using Eqs. (4.37), (4.43) and (4.46) give

$$\frac{dy}{d\varphi} = -4h(\varphi - \varphi_u) + o(\varphi - \varphi_u) \tag{4.55}$$

as $\varphi \to \varphi_u$ and

$$\frac{dy}{d\varphi} = 4h(\varphi - \varphi_l) + o(\varphi - \varphi_l) \tag{4.56}$$

as $\varphi \to \varphi_l$. Integrating Eq. (4.55) with the use of the boundary condition $y = h$ for $\varphi = \varphi_u$ and Eq. (4.56) with the use of the boundary condition $y = -h$ for $\varphi = \varphi_l$ yields

$$h - y = 2h(\varphi - \varphi_u)^2 + o\left[(\varphi - \varphi_u)^2\right], \tag{4.57}$$
$$y + h = 2h(\varphi - \varphi_l)^2 + o\left[(\varphi - \varphi_l)^2\right]$$

as $\varphi \to \varphi_u$ and $\varphi \to \varphi_l$, respectively. Substituting Eq. (4.57) into Eqs. (4.52) and (4.53) gives

$$\xi_{eq} = \sqrt{\frac{2}{3}} \frac{V}{\sqrt{h}} \frac{\sqrt{1-c}}{\left(1 - c\sin^2 2\theta\right)\sqrt{h-y}} + o\left[\frac{1}{\sqrt{h-y}}\right], \tag{4.58}$$
$$\xi_{eq} = \sqrt{\frac{2}{3}} \frac{V}{\sqrt{h}} \frac{\sqrt{1-c}}{\left(1 - c\sin^2 2\theta\right)\sqrt{h+y}} + o\left[\frac{1}{\sqrt{h+y}}\right]$$

as $y \to h$ and $y \to -h$, respectively. In the case under consideration, $s = h - y$ in the vicinity of the friction surface $y = h$ and $s = h + y$ in the vicinity of the friction surface $y = -h$. Therefore, it is evident that Eq. (4.58) coincides with Eq. (4.31) at $m = 2$.

References

1. Alexandrov S, Jeng Y-R (2013) Singular rigid/plastic solutions in anisotropic plasticity under plane strain conditions. Cont Mech Therm 25:685–689
2. Alexandrov S, Mustafa Y (2014) The strain rate intensity factor in the plane strain compression of thin anisotropic metal strip. Meccanica 49:2901–2906
3. Collins IF, Meguid SA (1977) On the influence of hardening and anisotropy on the plane-strain compression of thin metal strip. Trans ASME J Appl Mech 44:271–278
4. Hill R (1950) The mathematical theory of plasticity. Clarendon Press, Oxford
5. Rice JR (1973) Plane strain slip line theory for anisotropic rigid/plastic materials. J Mech Phys Solids 21:63–74

Chapter 5
Numerical Method

This chapter is concerned with a numerical method for calculating the strain rate intensity factor in plane strain deformation of rigid perfectly plastic material. It is evident that standard commercial packages are not capable for supplying solutions that satisfy the exact asymptotic expansion (2.20). For example, using ABAQUS a ring upsetting process has been analyzed in [5]. All the finite element analyses presented in this paper failed to converge in the case of the maximum friction law. The extended finite element method [9] might provide solutions in which Eq. (2.20) is satisfied. However, to the best of author's knowledge, no attempt has been made to develop this numerical technique for calculating the strain rate intensity factor. Moreover, it is believed that the method of characteristics is the best choice for models described by hyperbolic systems of equations. In particular, general numerical schemes based on the method of characteristics are available for plane strain deformation of rigid perfectly plastic material [3, 4, 7, 8, 10, 11]. In order to calculate the strain rate intensity factor, it is just necessary to supplement these schemes with a corresponding procedure. To the best of authors knowledge, the only available procedure has been proposed in [2]. Most of the present chapter is based on this work.

5.1 The Strain Rate Intensity Factor in Characteristic Coordinates

The system of equations consisting of Eqs. (1.1), (1.2), (1.15), and (1.16) is hyperbolic [10]. The characteristics for the stresses and the velocities coincide and, therefore, there are only two distinct characteristic directions at a point. The characteristic directions are orthogonal. Let the two families of characteristics be labeled by the parameters α and β. In general, one or two families of characteristics can be straight.

© The Author(s) 2018
S. Alexandrov, *Singular Solutions in Plasticity*, SpringerBriefs
in Continuum Mechanics, DOI 10.1007/978-981-10-5227-9_5

Fig. 5.1 Cartesian (x, y) and characteristic (α, β) coordinates

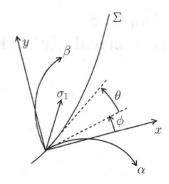

These special cases are excluded from consideration. The α- and β- lines are regarded as a pair of right-handed curvilinear orthogonal coordinates. By the convention, the orientation of these lines is chosen such that the algebraically greatest principal stress σ_1 falls in the first and third quadrants. Let ϕ be the anti-clockwise angular rotation of the α- line from the x- axis of a Cartesian coordinate system (x, y) (Fig. 5.1). Then [10],

$$\phi = \alpha + \beta, \tag{5.1}$$

$$\frac{1}{R} = \frac{\partial \phi}{\partial s_\alpha}, \quad \frac{1}{S} = -\frac{\partial \phi}{\partial s_\beta} \tag{5.2}$$

where $\partial/\partial s_\alpha$ and $\partial/\partial s_\beta$ are space derivatives taken along the α- and β- lines, respectively. Note that $\partial \phi/\partial s_\alpha = \partial \alpha/\partial s_\alpha$ since β is constant along the α- lines and $\partial \phi/\partial s_\beta = \partial \beta/\partial s_\beta$ since α is constant along the β- lines. Also, R is the radius of curvature of the α- lines and S is the radius of curvature of the β- lines. Those are algebraic quantities whose sign depends on the sense of the derivatives $\partial/\partial s_\alpha$ and $\partial/\partial s_\beta$. Let θ be the anti-clockwise rotation from the α- direction to the tangent to an arbitrary curve, Σ (Fig. 5.1). Then, using Eqs. (5.1) and (5.2) the space derivative along Σ is determined as

$$\frac{\partial}{\partial \Sigma} = \frac{\cos \theta}{R} \frac{\partial}{\partial \alpha} - \frac{\sin \theta}{S} \frac{\partial}{\partial \beta}. \tag{5.3}$$

Equation (1.3) is equivalent to

$$\xi_{eq} = \sqrt{\frac{2}{3}} \sqrt{\xi_{\zeta\zeta}^2 + \xi_{\eta\eta}^2 + 2\xi_{\zeta\eta}^2} \tag{5.4}$$

where (ζ, η) is any orthogonal coordinate system in the plane of flow. It is known that the normal strain rates in the characteristic coordinates vanish [10]. Therefore, it follows from Eq. (5.4) that

$$\xi_{eq} = \frac{2}{\sqrt{3}} |\xi_{\alpha\beta}| \tag{5.5}$$

where $\xi_{\alpha\beta}$ is the shear strain rate in the characteristic coordinates. On the other hand, $\xi_{\alpha\beta}$ is equal to the shear strain rate in the Cartesian coordinates, ξ_{xy}, at some point P if $\phi = 0$ at this point (Fig. 5.1). In this coordinate system, Eq. (1.2) for the shear strain rate becomes

$$\xi_{xy} = \frac{1}{2} \left(\frac{\partial u_x}{\partial y} + \frac{\partial u_y}{\partial x} \right) \tag{5.6}$$

where u_x and u_y are the Cartesian components of velocity. It is evident that (Fig. 5.1)

$$u_x = u_\alpha \cos\phi - u_\beta \sin\phi, \quad u_y = u_\alpha \sin\phi + u_\beta \cos\phi \tag{5.7}$$

where u_α and u_β are the velocity components referred to the α- and β- lines, respectively. Substituting Eq. (5.7) into Eq. (5.6) and putting $\phi = 0$ yield

$$2\xi_{xy} = \frac{\partial u_\alpha}{\partial y} - u_\beta \frac{\partial \phi}{\partial y} + u_\alpha \frac{\partial \phi}{\partial x} + \frac{\partial u_\beta}{\partial x} \tag{5.8}$$

at P. Assume that the curve Σ coincides with the x- axis. Then, $\theta = 0$ and it follows from Eq. (5.3) that

$$\frac{\partial}{\partial x} = \frac{\partial}{R\partial\alpha} \tag{5.9}$$

at P. Analogously, if the curve Σ coincides with the y- axis, then $\theta = \pi/2$ and it follows from Eq. (5.3) that

$$\frac{\partial}{\partial y} = -\frac{\partial}{S\partial\beta} \tag{5.10}$$

at P. Combining Eqs. (5.8) (5.10) and using Eq. (5.1) give

$$2\xi_{\alpha\beta} = -\frac{\partial u_\alpha}{S\partial\beta} + \frac{u_\beta}{S} + \frac{u_\alpha}{R} + \frac{\partial u_\beta}{R\partial\alpha}. \tag{5.11}$$

Substituting Eq. (5.11) into Eq. (5.5) results in

$$\xi_{eq} = \frac{1}{\sqrt{3}} \left| -\frac{\partial u_\alpha}{S\partial\beta} + \frac{u_\beta}{S} + \frac{u_\alpha}{R} + \frac{\partial u_\beta}{R\partial\alpha} \right|. \tag{5.12}$$

The expansion (2.20) is valid in the vicinity of an envelope of characteristics where $S = 0$ or $R = 0$. Consider the case $S = 0$. In the vicinity of a generic point Q on such a curve Eq. (5.12) becomes

$$\xi_{eq} = \frac{E_S}{|S|}, \quad E_S = \frac{1}{\sqrt{3}} \left| -\frac{\partial u_\alpha}{\partial\beta} + u_\beta \right| \tag{5.13}$$

to leading order. Here the derivative $\partial u_\alpha / \partial \beta$ and the velocity component u_β are understood to be calculated at the point Q. Comparing Eqs. (2.20) and (5.13) shows that

$$S = S_0 \sqrt{s} + o\left(\sqrt{s}\right) \tag{5.14}$$

as $s \to 0$. The tangent to the envelope under consideration coincides with the tangent to an α- line at Q. Therefore, Eq. (5.10) is valid with $s = my$ where $m = \pm 1$. Substituting Eq. (5.14) into this equation gives

$$\frac{\partial S}{\partial \beta} = -\frac{S_0^2}{2m} + o\,(1) \tag{5.15}$$

as $s \to 0$. Integrating Eq. (5.15) yields

$$S = -\frac{S_0^2}{2m} \left(\beta - \beta_Q\right) + o\left(\beta - \beta_Q\right) \tag{5.16}$$

as $\beta \to \beta_Q$. Here β_Q is the value of β at the point Q. Substituting Eq. (5.14) into Eq. (5.13) and comparing the resulting expression with Eq. (2.20) show that

$$D = \frac{E_S}{|S_0|}. \tag{5.17}$$

The dependence of u_α, u_β and S on α and β is supposed to be known from the solution of a boundary value problem. Therefore, the values of E_S and S_0 at Q can be found using the definition for E_S given in Eq. (5.13) and the asymptotic expansion given in Eq. (5.15) or (5.16). Then, the value of the strain rate intensity factor is readily determined from Eq. (5.17).

Consider an envelope of characteristics where $R = 0$. In the vicinity of a generic point Q on such a curve Eq. (5.12) becomes

$$\xi_{eq} = \frac{E_R}{|R|}, \quad E_R = \frac{1}{\sqrt{3}} \left| \frac{\partial u_\beta}{\partial \alpha} + u_\alpha \right| \tag{5.18}$$

to leading order. Here the derivative $\partial u_\beta / \partial \alpha$ and the velocity component u_α are understood to be calculated at the point Q. Comparing Eqs. (2.20) and (5.18) shows that

$$R = R_0 \sqrt{s} + o\left(\sqrt{s}\right) \tag{5.19}$$

as $s \to 0$. The tangent to the envelope under consideration coincides with the tangent to a β- line at Q. Therefore, Eq. (5.9) is valid with $s = mx$ where $m = \pm 1$. Substituting Eq. (5.19) into this equation gives

$$\frac{\partial R}{\partial \alpha} = \frac{R_0^2}{2m} + o\,(1) \tag{5.20}$$

as $s \to 0$. Integrating Eq. (5.20) yields

$$R = \frac{R_0^2}{2m}\left(\alpha - \alpha_Q\right) + o\left(\alpha - \alpha_Q\right) \tag{5.21}$$

as $\alpha \to \alpha_Q$. Here α_Q is the value α of at the point Q. Substituting Eq. (5.19) into Eq. (5.18) and comparing the resulting expression with Eq. (2.20) show that

$$D = \frac{E_R}{|R_0|}. \tag{5.22}$$

The dependence of u_α, u_β and R on α and β is supposed to be known from the solution of a boundary value problem. Therefore, the values of E_R and R_0 at Q can be found using the definition for E_R given in Eq. (5.18) and the asymptotic expansion given in Eq. (5.20) or (5.21). Then, the value of the strain rate intensity factor is readily determined from Eq. (5.22).

5.2   Benchmark Problem

In this section Eqs. (5.17) and (5.22) are used to determine the strain rate intensity factor by means of the approximate solution given in Sect. 2.3. Note that using this solution in Cartesian coordinates the strain rate intensity factor has been calculated in [1]. As a result,

$$D = \sqrt{\frac{2}{3}}\frac{V}{\sqrt{h}} \tag{5.23}$$

It is seen from this equation that the strain rate intensity factor is independent of x. Therefore, using the procedure described in Sect. 5.1 it is sufficient to find its value at one point of each of the maximum friction surfaces. Since the angle between the principal stress axis corresponding to σ_1 and the tangent to α- characteristic lines is equal to $\pi/4$, it is seen from Fig. 5.1 that

$$\psi = \phi + \frac{\pi}{4}. \tag{5.24}$$

Here ψ is the angle between the principal stress direction corresponding to σ_1 and x- axis measured from the axis anticlockwise. The velocity components u_α and u_β are related to u_x and u_y as (Fig. 5.1)

$$u_\alpha = u_x \cos\phi + u_y \sin\phi, \quad u_\beta = -u_x \sin\phi + u_y \cos\phi. \tag{5.25}$$

Fig. 5.2 Shape of
characteristics in the
approximate solution

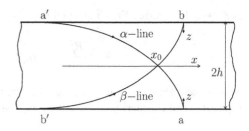

Using Eq. (5.24) it is possible to transform Eq. (2.61) to

$$\frac{\sigma_{xx}}{k_0} = \frac{\sigma}{k_0} - \sin 2\phi, \quad \frac{\sigma_{yy}}{k_0} = \frac{\sigma}{k_0} + \sin 2\phi, \quad \frac{\sigma_{xy}}{k_0} = \cos 2\phi. \tag{5.26}$$

Equations (2.69) and (5.24) combine to give

$$y = h \cos 2\phi. \tag{5.27}$$

Since $\sigma_{xx} - \sigma_{yy} \geq 0$, it is evident from Eq. (5.26) that

$$-\pi/2 \leq \phi \leq 0. \tag{5.28}$$

The equation for α- lines is [10]

$$\frac{dy}{dx} = \tan \phi. \tag{5.29}$$

Substituting Eq. (5.27) into Eq. (5.29) yields

$$\cos^2 \phi \frac{d\phi}{dx} = -\frac{1}{4h}. \tag{5.30}$$

Consider the α- line aa' through the point $(x_0, 0)$, as shown in Fig. 5.2. It is seen
from Eqs. (5.27) and (5.28) that $\phi = -\pi/4$ at $y = 0$. Integrating Eq. (5.30) and
using the boundary condition $\phi = -\pi/4$ for $x = x_0$ give

$$x = x_0 - h \left(2\phi + \sin 2\phi + \frac{\pi}{2} + 1 \right). \tag{5.31}$$

Equations (5.27) and (5.31) determine the α- line aa' in parametric form with ϕ
being the parameter varying in the range (5.28).
    The equation for β- lines is [10]

$$\frac{dy}{dx} = -\cot \phi. \tag{5.32}$$

Substituting Eq. (5.27) into Eq. (5.32) yields

$$\sin^2 \phi \frac{d\phi}{dx} = \frac{1}{4h}. \tag{5.33}$$

Consider the β- line bb' through the same point $(x_0, 0)$, as shown in Fig. 5.2. Integrating Eq. (5.33) and using the boundary condition $\phi = -\pi/4$ for $x = x_0$ give

$$x = x_0 + h \left(2\phi - \sin 2\phi + \frac{\pi}{2} - 1 \right). \tag{5.34}$$

Equations (5.27) and (5.34) determine the β- line bb' in parametric form with ϕ being the parameter varying in the range (5.28).

The magnitude of the curvature of a plane curve given in parametric form can be found as

$$|\kappa| = \frac{|\dot{x}\ddot{y} - \dot{y}\ddot{x}|}{[(\dot{x})^2 + (\dot{y})^2]^{3/2}} \tag{5.35}$$

where the superimposed dot denotes differentiation with respect to the parameter. Substituting Eqs. (5.27) and (5.31) into Eq. (5.35) and taking into account that $R \leq 0$ according to the convention adopted (see [10]) and that ϕ varies in the range (5.28) yield

$$R = -4h \cos \phi. \tag{5.36}$$

It is evident from this equation and Eq. (5.27) that $R = 0$ at $\phi = -\pi/2$ and, therefore, at the maximum friction surface $y = -h$. Expanding the right hand side of Eq. (5.36) in a series leads to

$$R = -4h \left(\phi + \frac{\pi}{2} \right) + o \left(\phi + \frac{\pi}{2} \right) \tag{5.37}$$

as $\phi \to -\pi/2$. Comparing Eqs. (5.21) and (5.37) with the use of Eq. (5.1) shows that $m = -1$ and

$$R_0 = 2\sqrt{2h}. \tag{5.38}$$

Using Eqs. (2.73), (2.78), (5.24), (5.25), (5.27) and (5.31) and taking into account Eq. (5.28) the variation of the velocity component u_β along the line aa' is determined as

$$\frac{u_\beta}{V} = - \left(a + \frac{x_0}{h} - \frac{\pi}{2} - 1 \right) \sin \phi + 2\phi \sin \phi - \cos \phi. \tag{5.39}$$

It follows from Eq. (5.1) that $\partial/\partial\phi = \partial/\partial\alpha$ along α- lines. Therefore, differentiating Eq. (5.39) with respect to ϕ and then putting $\phi = -\pi/2$ give

$$\frac{1}{V} \frac{\partial u_\beta}{\partial \alpha} = -3. \tag{5.40}$$

It is evident from Eq. (5.25) that $u_\alpha = -u_y$ at $\phi = -\pi/2$ and, then, from the boundary condition $u_y = V$ at $y = -h$ (see Fig. 2.2) that $u_\alpha = -V$ at $\phi = -\pi/2$. Substituting this value of u_α and the derivative $\partial u_\beta/\partial \alpha$ from Eq. (5.40) into Eq. (5.18) gives $E_R = 4V/\sqrt{3}$. Then, it follows from Eqs. (5.22) and (5.38) that the strain rate intensity factor is given by Eq. (5.23).

Substituting Eqs. (5.27) and (5.34) into Eq. (5.35) and taking into account that $S \leq 0$ according to the convention adopted (see [10]) and that ϕ varies in the range (5.28) yield

$$S = 4h \sin \phi. \tag{5.41}$$

It is evident from this equation and Eq. (5.27) that $S = 0$ at $\phi = 0$ and, therefore, at the maximum friction surface $y = h$. Expanding the right hand side of Eq. (5.41) in a series gives

$$S = 4h\phi + o(\phi) \tag{5.42}$$

as $\phi \to 0$. Comparing Eqs. (5.15) and (5.42) with the use of Eq. (5.1) shows that $m = -1$ and

$$S_0 = 2\sqrt{2h}. \tag{5.43}$$

Using Eqs. (2.73), (2.78), (5.24), (5.25), (5.27) and (5.34) and taking into account Eq. (5.28) the variation of the velocity component u_α along the line bb' is determined as

$$\frac{u_\alpha}{V} = \left(a + \frac{x_0}{h} + \frac{\pi}{2} - 1\right) \cos \phi + 2\phi \cos \phi + \sin \phi. \tag{5.44}$$

It follows from Eq. (5.1) that $\partial/\partial \phi = \partial/\partial \beta$ along β- lines. Therefore, differentiating Eq. (5.44) with respect to ϕ and then putting $\phi = 0$ give

$$\frac{1}{V}\frac{\partial u_\alpha}{\partial \beta} = 3. \tag{5.45}$$

It is evident from Eq. (5.25) that $u_\beta = u_y$ at $\phi = 0$ and, then, from the boundary condition (2.52) that $u_\beta = -V$ at $\phi = 0$. Substituting this value of u_β and the derivative $\partial u_\alpha/\partial \beta$ from Eq. (5.45) into Eq. (5.13) gives $E_S = 4V/\sqrt{3}$. Then, it follows from Eqs. (5.17) and (5.43) that the strain rate intensity factor is given by Eq. (5.23).

5.3 Numerical Solution

5.3.1 General Numerical Scheme for the Boundary Value Problem of Section 2.3

The approximate boundary conditions (2.53) and (2.55) are not used in this section. The boundary conditions consist of Eqs. (2.51), (2.52), (2.54), (2.64) and

$$u_x = 0 \tag{5.46}$$

for $x = w$,

$$\sigma_{xx} = 0 \tag{5.47}$$

for $x = 0$, and

$$\sigma_{xy} = 0 \tag{5.48}$$

for $x = 0$. The numerical scheme developed in [11] is supplemented with a numerical procedure to determine the strain rate intensity factor according to Eqs. (5.17) and (5.22). Then, the strain rate intensity factor in compression of a finite layer between two rough plates is found. The numerical solution for stress and velocity in this process has been obtained in [11] assuming the maximum friction law and in [6] assuming that the frictional shear traction on the plate faces is a constant proportion of the shear yield stress. Therefore, the main contribution of the present solution is the distribution of the strain rate intensity factor along the friction surface.

The general structure of the characteristic field found in [11] and Cartesian coordinates (x, y) are illustrated in Fig. 5.3. Both families of characteristics are straight in the region AO_1B, α- lines are straight in the region AB_1O_1, and β- lines are straight in the region BA_1O_1. No numerical treatment is required in these regions. In particular, O_1B_1 and O_1A_1 are circular arcs with their centers at the points A and B, respectively. Therefore,

$$S = -\sqrt{2}h \tag{5.49}$$

on O_1B_1 and

$$R = -\sqrt{2}h \tag{5.50}$$

on O_1A_1 by simple geometry. The sense of R and S has been chosen according to Eq. (5.2). The equations for the radii of curvature are [10]

$$dS + Rd\phi = 0 \quad \text{along} \quad \text{an} \quad \alpha\text{- line} \tag{5.51}$$
$$dR - Sd\phi = 0 \quad \text{along} \quad \text{a} \quad \beta\text{- line}$$

In the problem under consideration, the β- lines are orthogonal to the maximum friction surface $y = h$, and the α- lines to the maximum friction surface $y = -h$.

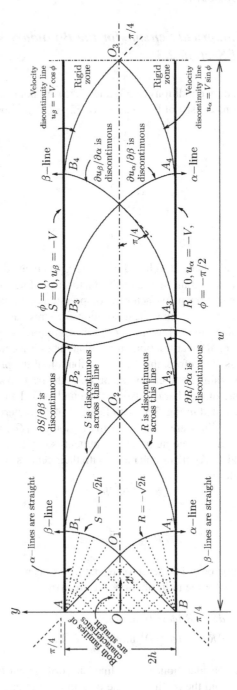

Fig. 5.3 General structure of the characteristic field found in [11]

Therefore,

$$S = 0 \tag{5.52}$$

at $y = h$ and $x \geq \sqrt{2}h$ and

$$R = 0 \tag{5.53}$$

at $y = -h$ and $x \geq \sqrt{2}h$. The choice of the coordinate system and the definition for ϕ dictate that $\phi = 0$ at $y = h$, $\phi = -\pi/4$ at the axis of symmetry $y = 0$ and $\phi = -\pi/2$ at $y = -h$. Using the boundary conditions (5.49), (5.50), (5.52) and (5.53) Eq. (5.51) has been solved numerically in the region $O_1 B_1 B_4 O_3 A_4 A_1 O_1$ by the method described in [11]. Note that $d\phi = d\alpha$ along the α- lines and $d\phi = d\beta$ along the β- lines, as follows from Eq. (5.1). The characteristic lines $B_4 O_3$ and $A_4 O_3$ are determined by the condition that they intersect at the center of the layer. These curves are rigid plastic boundaries. The velocity vector is discontinuous across these curves. The rigid zones are moving toward each other with speed V. Since the normal velocity must be continuous, its value on the plastic side of the rigid plastic boundaries can be found with the use of the definition for ϕ as

$$u_\alpha = V \sin \phi \tag{5.54}$$

on $O_3 A_4$ and

$$u_\beta = -V \cos \phi \tag{5.55}$$

on $O_3 B_4$. It has been taken into account here that $O_3 B_4$ is an α- line and, therefore, u_β is directed along the normal to this line, and $O_3 A_4$ is an β- line and, therefore, u_α is directed along the normal to this line. The equations for u_α and u_β are [10]

$$du_\alpha - u_\beta d\phi = 0 \quad \text{along an } \alpha\text{- line} \tag{5.56}$$
$$du_\beta + u_\alpha d\phi = 0 \quad \text{along a } \beta\text{- line.}$$

Since the β- lines are orthogonal to the surface $y = h$, and the α- lines to the surface $y = -h$,

$$u_\beta = -V \tag{5.57}$$

at $y = h$ and

$$u_\alpha = -V \tag{5.58}$$

at $y = -h$ (Fig. 5.3). Using the boundary conditions (5.54), (5.55), (5.57) and (5.58) together with the method presented in [11] Eq. (5.56) has been solved numerically in the region $O_1 B_1 B_4 O_3 A_4 A_1 O_1$.

The method developed in [11] has been supplemented with an additional procedure for calculating the strain rate intensity factor. To describe this procedure, it is sufficient to consider a generic point P on the maximum friction surface $y = h$. A discrete network of characteristic curves in the vicinity of this point is shown in

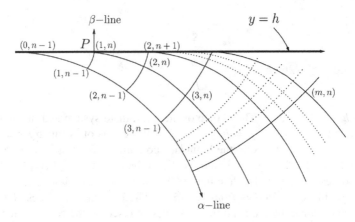

Fig. 5.4 Discrete mesh of characteristic coordinates near point P

Fig. 5.4. Suppose that the positions of the nodal points of the α- line through the point $(0, \ n-1)$ along with the values of R and S at these nodal points are known. Consider the β- line through the point P. To maintain a second order accuracy, the curvature S for the arc $(1, \ n-1)$ to $(1, \ n)$ is represented as

$$S = S_{(1, n-1)} + S'\left(\phi - \phi_{(1, n-1)}\right) + S''\left(\phi - \phi_{(1, n-1)}\right)^2 + O\left[\left(\phi - \phi_{(1, n-1)}\right)^3\right] \quad (5.59)$$

as $\phi \to \phi_{(1, \ n-1)}$. Here $S_{(1, \ n-1)}$ and $\phi_{(1, \ n-1)}$ are the values of S and ϕ at $(1, \ n-1)$. These quantities are known. S' and S'' are unknown and should be found from the conditions that $S = 0$ and $y = h$ at the point $(1, \ n)$. Substituting the former into Eq. (5.59) gives

$$S_{(1,n-1)} - S'\phi_{(1,n-1)} + S''\phi_{(1,n-1)}^2 = 0. \quad (5.60)$$

It has been taken into account here that $\phi = 0$ at the point $(1, \ n)$ (Fig. 5.3). Using Eqs. (5.1) and (5.2) and the definition for ϕ the variation of x and y along β- lines is given by $dx = -ds_\beta \sin \phi = S \sin \phi d\phi$ and $dy = ds_\beta \cos \phi = -S \cos \phi d\phi$. For the arc $(1, \ n-1)$ to $(1, \ n)$ these equations are approximated as

$$x_{(1, n)} - x_{(1, n-1)} = \int_{\phi_{(1, n-1)}}^{0} S \sin \phi d\phi, \quad h - y_{(1, n-1)} = -\int_{\phi_{(1, n-1)}}^{0} S \cos \phi d\phi \quad (5.61)$$

where $x_{(1, \ n-1)}$ and $x_{(1, \ n)}$ are the x- coordinates of the points $(1, \ n-1)$ and $(1, \ n)$, respectively. The former is known. The radius of curvature S in the integrands in Eq. (5.61) should be eliminated by means of Eq. (5.59). Equations (5.60) and (5.61)[2] constitute the system of equations for determining S' and S''. Having evaluated S' and S'' the x- coordinate of the point $(1, \ n)$ is found from Eq. (5.61)[1]. It is evident from Eq. (5.1) that $\partial/\partial\beta = \partial/\partial\phi$ along a β- line. Then, comparing Eqs. (5.15) and (5.59) yields

$$S_0 = \sqrt{2\left|2S''\phi_{(1,n-1)} - S'\right|}. \tag{5.62}$$

Having calculated the velocity field the derivative $\partial u_\alpha/\partial\beta$ at P has been approximated as $\partial u_\alpha/\partial\beta \approx \left(u_{\alpha(1,n)} - u_{\alpha(1,n-1)}\right)/\left(\beta_{(1,n)} - \beta_{(1,n-1)}\right)$. It is evident that $u_\beta = -V$ at P. Then, using Eq. (5.62) the strain rate intensity factor is determined by means of Eq. (5.17).

5.3.2 Verification of the Accuracy of the Numerical Scheme

In order to verify the accuracy of the strain rate intensity factor found by means of the numerical scheme presented in Sect. 5.3.1, the strain rate intensity factor given by Eq. (5.23) is approximated numerically assuming that $h = 1$ and $V = 1$. To this end, it is first necessary to replace the circular arc $O_1 A_1$ (Fig. 5.3) with the curve given by Eqs. (5.27) and (5.31), and the circular arc $O_1 B_1$ with the curve given by Eqs. (5.27) and (5.34). The value of x_0 involved in Eqs. (5.31) and (5.34) is not essential for determining the strain rate intensity factor. Starting from these baselines a network of characteristics is calculated including the value of S_0. The distribution of R and S is illustrated in Fig. 5.5. It is seen from this figure that it is practically independent of x. It follows from Eq. (2.77) that the velocity u_x is a linear function of x. Figure 5.6 demonstrates that the velocity u_α is a linear function of x at $y = 0$ and $y = h = 1$. It is evident from Eq. (5.25) that $u_x = u_\alpha$ at $y = h$ (or $\phi = 0$) and $u_x = \sqrt{2}u_\alpha = \sqrt{2}u_\beta$ at $y = 0$ (or $\phi = -\pi/4$). Therefore, the numerical solution also predicts that u_x is a linear function of x. Having calculated S_0 and the velocity field the strain rate intensity factor is found by means of Eq. (5.17). Since it depends on V and h it is convenient to introduce the dimensionless strain rate intensity factor by $d = D\sqrt{h}/V$. The variation of d with x is shown in Fig. 5.7 by the broken line. The solid line corresponds to the exact value given by Eq. (5.23). Introduce the relative error by $\delta = |d_e - d_n|/d_e$ where d_e is the exact value of d from Eq. (5.23) and d_n is the value of d from the numerical solution. The variation of δ with x is also depicted in Fig. 5.7. It is seen from this figure that the value of the relative error is very small. The error grows linearly from the right towards the left and its maximum occurs at the baseline. This is associated with the error accumulation due to the numerical marching scheme. The maximum relative error in the domain considered, $\delta_{\max}$, is about $3 \cdot 10^{-3}$. The mesh dependence of the scheme is estimated by the value of $\delta_{\max}$ with respect to the number of nodal points on the baselines, N. This dependence is shown in logarithmic scales in Fig. 5.8. The slope of the convergence curve suggests that the scheme has a first order accuracy for the strain rate intensity factor. Lowering of the scheme accuracy for the strain rate

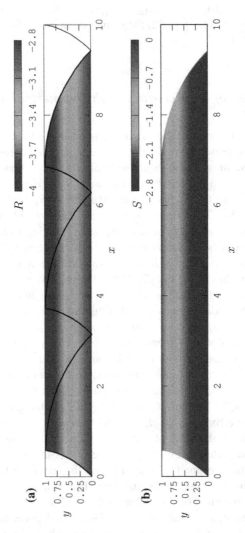

Fig. 5.5 Distribution of the radii of curvature of characteristic curves in the numerical approximation of the analytic solution presented in Sect. 5.2. The number of nodal points on the baselines is 270

Fig. 5.6 Distribution of the velocity component u_α along the lines $y = 0$ and $y = h = 1$. The number of nodal points on the baselines is 270

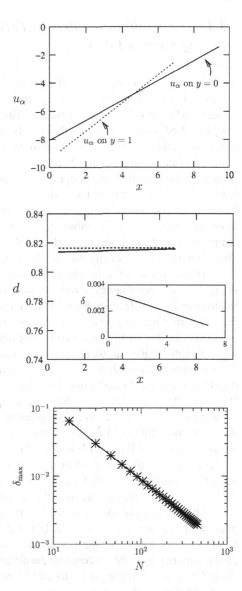

Fig. 5.7 Comparison of analytical and numerical values of the strain rate intensity factor and the distribution of the relative error along the friction surface. The number of nodal points on the baselines is 270

Fig. 5.8 Variation of the maximum relative error δ_{max} with N

intensity factor is associated with the derivative operation involved in Eq. (5.13). This is not considered as a drawback because the scheme does not have implicit iterations and the calculation is fast. The accuracy can be improved simply by increasing the number of nodal points on the baselines.

5.3.3 The Strain Rate Intensity Factor in Compression
of a Finite Layer

As in Sect. 5.3.2, it is assumed, with no loss of generality, that $h = 1$ and $V = 1$. The point O_2 (Fig. 5.3) coincides with the center of the layer, the point O_3, when the ratio of width to height is 3.64, approximately. Therefore, the solution is valid if and only if $w/h \geq 3.64$. Figure 5.9 shows the distribution of R, S, u_α and u_β at $w/h = 9.85$. One of the radii of curvature is discontinuous across the β- line between the points A_1 and B_2 and across the α- line between the points B_1 and A_2. On the other hand, the velocity field is continuous except the rigid plastic boundaries. Nevertheless, there are lines of discontinuity of the derivatives $\partial u_\alpha/\partial \beta$ and $\partial u_\beta/\partial \alpha$. Those are the β- line between the points A_3 and B_4 and the α- line between the points B_3 and A_4. There are also lines of discontinuity of the derivatives $\partial S/\partial \beta$ and $\partial R/\partial \alpha$. Those are the α- line through the point B_2 and the β- line through the point A_2. All the discontinuity lines are diagrammatically shown in Fig. 5.3.

The dependence of the dimensionless strain rate intensity factor on x is presented in Fig. 5.10 at several values of w/h. All the dashed lines correspond to the value of d found from Eq. (5.23). In contrast to this approximate solution ignoring end effects, the strain rate intensity factor for a finite layer varies along the friction surface. Moreover, the distribution of the strain rate intensity factor is discontinuous if w/h is large enough. In general, there are two sources of these discontinuities. One source is the centered fans $B_1 O_1 A$ and $A_1 O_1 B$ resulting in a discontinuity of the derivative $\partial S/\partial \beta$ at the point B_2 and a discontinuity of the derivative $\partial R/\partial \alpha$ at the point A_2 (Fig. 5.3). These derivatives are involved in Eqs. (5.17) and (5.22). The other source is the rigid zones resulting in a discontinuity of the derivative $\partial u_\alpha/\partial \beta$ at the point B_3 and a discontinuity of the derivative $\partial u_\beta/\partial \alpha$ at the point A_3. These derivatives are also involved in Eqs. (5.17) and (5.22). Let x_{B_2} be the x- coordinate of the point B_2 (and A_2 indeed). It is evident that the value of x_{B_2} is independent of w. The numerical solution shows that $x_{B_2} \approx 4.19\,h$ and that the points B_2 and B_4 coincide if $w/h \approx 6.72$. Therefore, this type of discontinuity in the strain rate intensity factor caused by the structure of the characteristic field in the vicinity of the free edge is possible if and only if $w/h > 6.72$. In Fig. 5.10, these discontinuities are indicated by open circles. Let x_{B_3} be the x- coordinate of the point B_3 (and A_3 indeed). In this case it is more reasonable to consider the distance $w - x_{B_3}$. It is evident that the points B_1 and B_3 coincide if the points B_2 and B_4 coincide, i.e. at $w/h \approx 6.72$. Therefore, the distribution of the strain rate intensity factor is continuous for $w/h < 6.72$. It is seen in Fig. 5.10. The distance $w - x_{B_3}$ slightly increases with w. In particular, $w - x_{B_3} \approx 3.09\,h$ at $w/h = 7.76$ and $w - x_{B_3} \approx 3.14\,h$ at $w/h = 16.14$. The discontinuities in the strain rate intensity factor caused by the existence of the rigid zones are indicated by closed circles in Fig. 5.10. The discontinuities of the two types coincide if $w/h \approx 9.85$. It is evident that $x_{B_2} > x_{B_3}$ if $w/h < 9.85$ and vice versa. The case $x_{B_2} < x_{B_3}$ at $w/h = 11.95$ is illustrated in Fig. 5.11a and the $x_{B_2} > x_{B_3}$

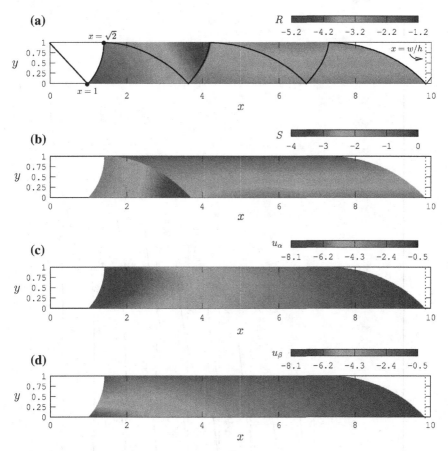

Fig. 5.9 Illustration of the numerical solution for R, S, u_α and u_β at $w/h = 9.85$. The number of nodal points on the baselines is 270

at $w/h = 7.76$ in Fig. 5.11b. The points of discontinuity are marked by squares. The discontinuities associated with the solid lines are caused by the structure of the characteristic field in the vicinity of the free edge and the discontinuities associated with the dashed lines by the existence of the rigid zones.

It is seen from Fig. 5.10 that the numerical value of the strain rate intensity factor in the region between the points of discontinuity is very close to that given by Eq. (5.23) if $w/h > 9.85$.

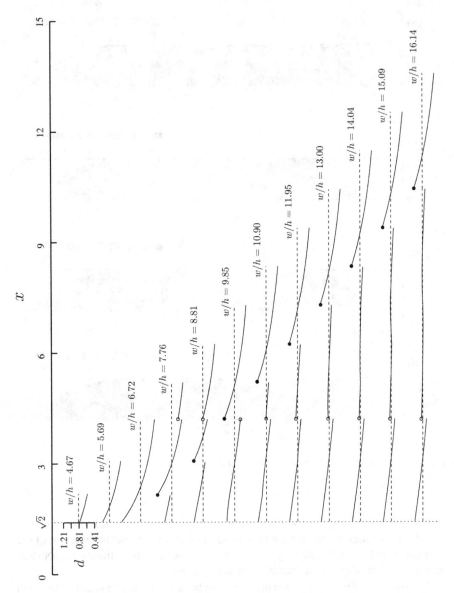

Fig. 5.10 Distribution of the strain rate intensity factor along the friction surface at several values of *w/h*

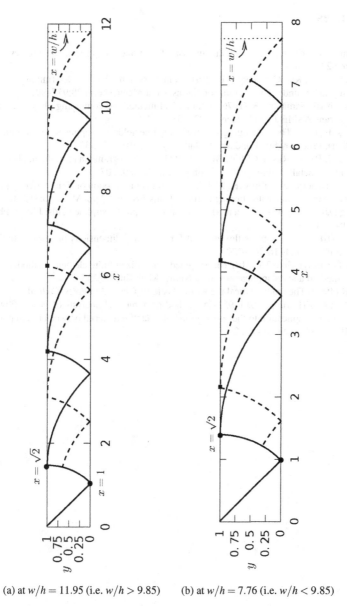

(a) at $w/h = 11.95$ (i.e. $w/h > 9.85$)   (b) at $w/h = 7.76$ (i.e. $w/h < 9.85$)

Fig. 5.11 Location of points of strain rate intensity factor discontinuity

References

1. Alexandrov S (2009) The strain rate intensity factor and its applications: a review. Mater Sci Forum 623:1–20
2. Alexandrov S, Kuo C-Y, Jeng Y-R (2016) A numerical method for determining the strain rate intensity factor under plane strain conditions. Cont Mech Therm 28:977–992
3. Bachrach BI, Samanta SK (1976) A numerical method for computing plane plastic slip-line fields. Trans ASME J Appl Mech 43:97–101
4. Collins IF (1968) The algebraic-geometry of slip line fields with applications to boundary value problems. Proc Roy Soc Lond Ser A Math Phys Sci 303:317–338
5. Chen J-C, Pan C, Rogue CMOL, Wang H-P (1998) A Lagrangian reproducing kernel particle method for metal forming analysis. Comp Mech 22:289–307
6. Das NS, Banerjee J, Collins IF (1979) Plane strain compression of rigid perfectly plastic strip between parallel dies with slipping friction. Trans ASME J Appl Mech 46:317–321
7. Ewing DJF (1967) A series-method for constructing plastic slipline fields. J Mech Phys Solids 15:105–114
8. Ewing DJF (1968) A mass-flux method for deducing dimensions of plastic slipline fields. J Mech Phys Solids 16:267–272
9. Fries T-P, Belytschko T (2010) The extended/generalized finite element method: an overview of the method and its applications. Int J Numer Meth Eng 84:253–304
10. Hill R (1950) The mathematical theory of plasticity. Clarendon Press, Oxford
11. Hill R, Lee EH, Tupper SJ (1951) A method of numerical analysis of plastic flow in plane strain and its application to the compression of a ductile material between rough plates. Trans ASME J Appl Mech 18:46–52

Chapter 6
Applications

6.1 Generation of Fine Grain Layers Near Frictional Interfaces

Numerous experimental observations demonstrate that a narrow layer with drastically modified microstructure is often generated near frictional interfaces. This layer is usually called white layer in papers devoted to machining processes and fine grain layer in papers devoted to metal forming processes. A complete review of results on white layer/fine grain layer generation published before 1987 has been presented in [21]. According to this review article there are three main contributory mechanisms responsible for white layer generation. One of them is the mechanism of intensive plastic deformation. This mechanism can be described by means of the strain rate intensity factor introduced in Eq. (2.20). Such models have been proposed in [6, 22]. A correlation between the strain rate intensity factor and the thickness of the fine grain layer generated in direct extrusion of AZ31 alloy has been demonstrated in [8]. Most of the present chapter is based on this work. Other results related to tests designed to generate a fine grain layer near frictional interfaces in metal forming processes have been provided in [27–29].

The nominal chemical composition of the AZ31 magnesium alloy used in [8] is shown in Table 6.1. A round rod of initial diameter D_2 is reduced to diameter D_1 by pushing through a conical die. The opening angle of the die is 2ϕ. The distance between the inlet and outlet boundaries is L (Fig. 6.1). The values of D_1 and L are fixed. In particular, $D_1 = 13$ mm and $L = 40$ mm. Three dies with $\phi = 5$ deg, $\phi = 10$ deg, $\phi = 15$ deg are used. Then, diameter D_2 is found as $D_2 = 2L \tan \phi + D_1$. Three nominally identical billets have been extruded through each die. The initial microstructure of material is illustrated in Fig. 6.2. It is seen from this figure that the distribution of average grain size is quite uniform. The microstructure of material developed in the vicinity of the friction surface and near the axis of symmetry in course of extrusion through the $\phi = 5$ deg die is illustrated in Fig. 6.3, in course of extrusion through the $\phi = 10$ deg die in Fig. 6.4, and in course of extrusion through the $\phi = 15$ deg die in Fig. 6.5. The layer of fine grains is clearly seen in these figures.

© The Author(s) 2018
S. Alexandrov, *Singular Solutions in Plasticity*, SpringerBriefs
in Continuum Mechanics, DOI 10.1007/978-981-10-5227-9_6

Table 6.1 Nominal chemical composition of AZ31 alloy (mass fraction, %)

Al	Ni	Cu	Mn	Si	Zn	Mg
2.4–3.6	> 0.03	> 0.1	< 0.15	> 0.1	0.5–1.5	Rem

Fig. 6.1 Schematic illustration of the extrusion die

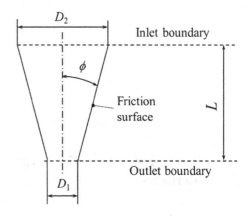

The radial distribution of the average grain size is shown in Fig. 6.6 for $\phi = 5$ deg, in Fig. 6.7 for $\phi = 10$ deg, and in Fig. 6.8 for $\phi = 15$ deg. Nano-indentation tests have been also performed using a Hysitron Triboscope with a Berkovich tip. Hardness has been measured at the same cross-sections where the microstructure was observed. The distribution of hardness is depicted in Fig. 6.9 for $\phi = 5$ deg, in Fig. 6.10 for $\phi = 10$ deg, and in Fig. 6.11 for $\phi = 15$ deg. There is an obvious correlation between the distributions of average grain size and hardness in Figs. 6.3, 6.4, 6.5, 6.6, 6.7, 6.8, 6.9, 6.10 and 6.11. The friction surface in these figures corresponds to $r = 6.5$ mm. In order to accurately determine the thickness of the fine grain layer, hardness has been measured at a larger number of points in the very vicinity of the friction surface. The distribution of hardness within the fine grain layer is illustrated in Fig. 6.12 for $\phi = 5$ deg, in Fig. 6.13 for $\phi = 10$ deg, and in Fig. 6.14 for $\phi = 15$ deg. The friction surface in these figures corresponds to the origin of the horizontal axis. Using the distributions of hardness shown in Figs. 6.12, 6.13 and 6.14 the thickness of the layer of fine grains has been evaluated. The dependence of the thickness of the layer as well as the average grain size and hardness in the layer on the die angle is shown in Table 6.2. It is seen from this table that the thickness of the layer and hardness in the layer increase whereas the average grain size in the layer decreases as the angle ϕ increases.

As it has been mentioned in Sect. 5, the only available numerical method for calculating the strain rate intensity factor is restricted to plane strain deformation of rigid perfectly plastic material. Therefore, the only possibility to find the strain rate intensity factor in axisymmetric extrusion is to use semi-analytical solutions. Such a solution has been provided in [34] where flow of rigid perfectly plastic material through an infinite converging conical channel is considered. It is believed that this

(a)   (b)

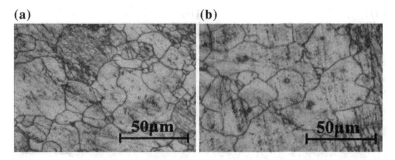

Fig. 6.2 Initial microscrusture of AZ31 alloy (billet material) at *center* (**a**) and surface (**b**)

(a)   (b)

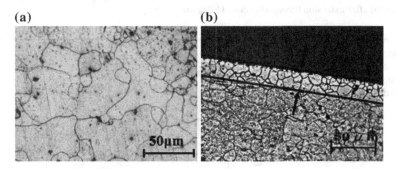

Fig. 6.3 Microstructure of extruded material at the axis of symmetry (**a**) and near the friction surface (**b**) after extrusion through the $\phi = 5$ deg die

(a)   (b)

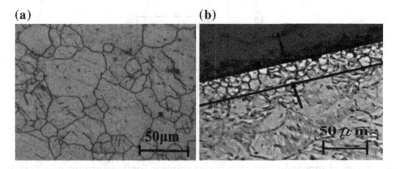

Fig. 6.4 Microstructure of extruded material at the axis of symmetry (**a**) and near the friction surface (**b**) after extrusion through the $\phi = 10$ deg die

Table 6.2 Dependence of parameters characterizing the fine grain layer generated near the friction surface on the angle ϕ

ϕ	5 deg	10 deg	15 deg
Thickness of layer, μm	26	32	36
Average grain size in layer, μm	7.5 ± 1.4	6.8 ± 1.8	5.5 ± 1.3
Average hardness in layer, GPa	1.88 ± 0.15	1.95 ± 0.15	2.12 ± 0.13

(a)   **(b)**

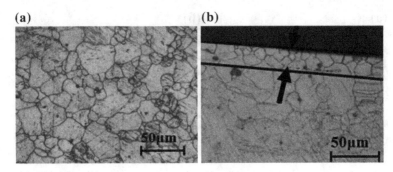

Fig. 6.5 Microstructure of extruded material at the axis of symmetry (**a**) and near the friction surface (**b**) after extrusion through the $\phi = 15$ deg die

Fig. 6.6 Distribution of average grain size along the radius of the sample after extrusion through the $\phi = 5$ deg die

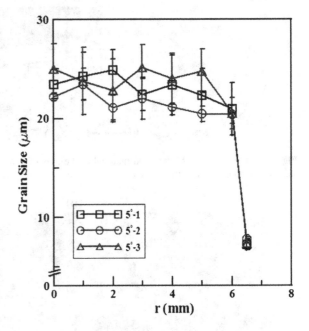

solution is a good approximation of the exact solution for flow through a finite channel if the latter is long enough. The solution for the infinite channel is given in a spherical coordinate system (ρ, θ, φ) taken relative to the axis of symmetry and the virtual apex of the channel. In particular, the surface of the channel is determined by the equation $\theta = \phi$. Assuming the von Mises yield criterion the solution of the boundary value problem reduces to the equation [34]

$$\frac{d\tau}{d\theta} + \tau \cot \theta + 2\sqrt{3}\sqrt{1 - \tau^2} = c. \qquad (6.1)$$

Fig. 6.7 Distribution of average grain size along the radius of the sample after extrusion through the $\phi = 10$ deg die

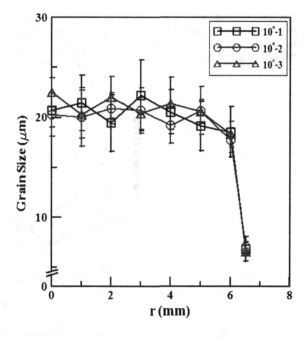

Fig. 6.8 Distribution of average grain size along the radius of the sample after extrusion through the $\phi = 15$ deg die

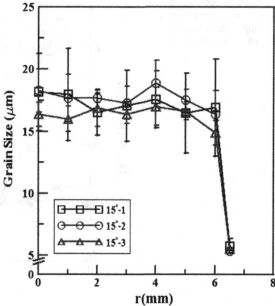

Fig. 6.9 Distribution of
hardness along the radius of
the sample after extrusion
through the $\phi = 5$ deg die

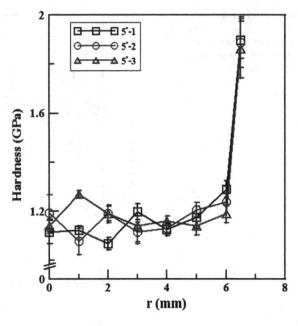

Fig. 6.10 Distribution of
hardness along the radius of
the sample after extrusion
through the $\phi = 10$ deg die

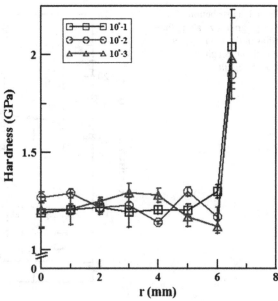

Fig. 6.11 Distribution of hardness along the radius of the sample after extrusion through the $\phi = 15$ deg die

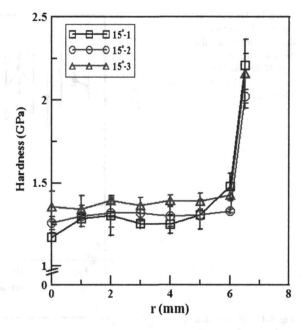

Fig. 6.12 Distribution of hardness in the vicinity of the friction surface after extrusion through the $\phi = 5$ deg die. The thickness of the hard layer is about 26 μm

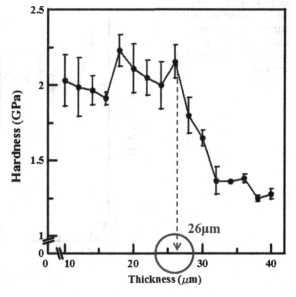

Fig. 6.13 Distribution of hardness in the vicinity of the friction surface after extrusion through the $\phi = 10$ deg die. The thickness of the hard layer is about 32 μm

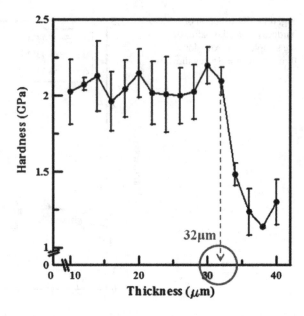

Fig. 6.14 Distribution of hardness in the vicinity of the friction surface after extrusion through the $\phi = 15$ deg die. The thickness of the hard layer is about 36 μm

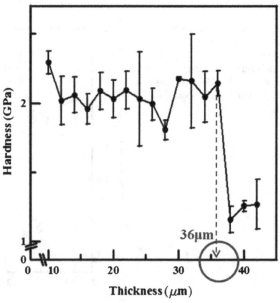

Here $\tau = \sigma_{\rho\theta}/k_0$, $\sigma_{\rho\theta}$ is the shear stress in the spherical coordinate system and c is a constant of integration. Symmetry demands

$$\tau = 0 \tag{6.2}$$

for $\theta = 0$. The maximum friction law (1.28) is equivalent to

$$\tau = 1 \tag{6.3}$$

for $\theta = \phi$. It is convenient to use the following substitution:

$$\tau = \cos \gamma. \tag{6.4}$$

Substituting Eq. (6.4) into Eqs. (6.1)–(6.3) and considering γ as an independent variable result in

$$\frac{d\theta}{d\gamma} = \frac{\sin \gamma \tan \theta}{\cos \gamma + \left(2\sqrt{3}\sin \gamma - c\right)\tan \theta}, \tag{6.5}$$

$$\theta = 0 \tag{6.6}$$

for $\gamma = \pi/2$ and

$$\theta = \phi \tag{6.7}$$

for $\gamma = 0$. Solving Eq. (6.5) numerically and satisfying the boundary conditions (6.6) and (6.7) determines c for a given value of ϕ. In particular, $c \approx 25.3$ for $\phi = 5$ deg, $c \approx 13.9$ for $\phi = 10$ deg, and $c \approx 10.1$ for $\phi = 15$ deg. The circumferential velocity vanishes everywhere. The radial velocity is given by [34]

$$u_\rho = -\frac{B}{\rho^2} \exp\left[-2\sqrt{3} \int_\phi^\theta \frac{\tau}{\sqrt{1-\tau^2}} d\theta\right] \tag{6.8}$$

where B is a constant of integration. Using Eqs. (6.4) and (6.5) it is possible to transform Eq. (6.8) to

$$u_\rho = -\frac{B}{\rho^2} \exp\left[-2\sqrt{3} \int_0^\gamma \frac{\cos \chi \tan \theta}{\cos \chi + \left(2\sqrt{3}\sin \chi - c\right)\tan \theta} d\chi\right]. \tag{6.9}$$

Here, χ is a dummy variable of integration and θ in the integrand should be replaced with the solution of Eq. (6.5). Let Q be the flux of material. Then,

$$Q = -2\pi \int_0^\phi \rho^2 u_\rho \sin\theta d\theta. \tag{6.10}$$

Eliminating the radial velocity in this equation by means of Eq. (6.9) and replacing integration with respect to θ with integration with respect to γ by means of Eq. (6.5) lead to the following equation for B

$$Q = -2\pi B \int_0^{\pi/2} \frac{\exp\left[-Q_1\left(\gamma,\theta\right)\right]\sin\gamma\tan\theta\sin\theta}{\left[\cos\gamma + \left(2\sqrt{3}\sin\gamma - c\right)\tan\theta\right]}d\gamma, \tag{6.11}$$

$$Q_1\left(\gamma,\theta\right) = 2\sqrt{3}\int_0^\gamma \frac{\cos\chi\tan\theta}{\cos\chi + \left(2\sqrt{3}\sin\chi - c\right)\tan\theta}d\chi.$$

It is worthy of note here that the values of c for the angles ϕ used in the experiment have been already calculated. Since the flux of material is known, Eq. (6.11) enables B to be found by integration. The shear strain rate in the spherical coordinate system is determined from Eq. (6.8) with the use of Eq. (6.4) as

$$\xi_{\rho\theta} = \frac{1}{2\rho}\frac{\partial u_\rho}{\partial\theta} = -\frac{\sqrt{3}}{\rho}\frac{\tau u_\rho}{\sqrt{1-\tau^2}} = -\frac{\sqrt{3}u_\rho\cot\gamma}{\rho}. \tag{6.12}$$

It is evident that this strain rate component approaches infinity as $\gamma \to 0$. Since the other strain components are bounded, $\xi_{eq}/\xi_{\rho\theta} \to 2/\sqrt{3}$ as $\gamma \to 0$. Therefore, it follows from Eq. (2.20) that

$$\xi_{\rho\theta} = \frac{\sqrt{3}}{2}\frac{D}{\sqrt{s}} + o\left(\frac{1}{\sqrt{s}}\right) \tag{6.13}$$

as $s \to 0$. On the other hand, Eq. (6.12) can be represented as

$$\xi_{\rho\theta} = \frac{\sqrt{3}B}{\rho^3\gamma} + o\left(\frac{1}{\gamma}\right) \tag{6.14}$$

as $\gamma \to 0$. It has been taken into account here that $u_\rho = -B/\rho^2$ at $\gamma = 0$, as follows from Eq. (6.9). In the vicinity of the surface $\gamma = 0$ Eq. (6.5) becomes

$$\frac{d\theta}{d\gamma} = \frac{\gamma\tan\phi}{1 - c\tan\phi} + o\left(\gamma\right) \tag{6.15}$$

as $\gamma \to 0$. Integrating Eq. (6.15) and using the boundary condition (6.7) result in

$$\phi - \theta = \frac{\tan \phi}{2 \, (c \tan \phi - 1)} \gamma^2 + o \left(\gamma^2 \right) \tag{6.16}$$

as $\gamma \to 0$. Substituting Eq. (6.16) into Eq. (6.14) yields

$$\xi_{\rho\theta} = \sqrt{\frac{3 \tan \phi}{2 \, (c \tan \phi - 1)} \frac{B}{\rho^3 \sqrt{\phi - \theta}}} + o \left(\frac{1}{\sqrt{\phi - \theta}} \right) \tag{6.17}$$

as $\theta \to \phi$. Comparing Eqs. (6.13) and (6.17) shows that

$$D = \frac{B}{\rho^{5/2}} \sqrt{\frac{2 \tan \phi}{c \tan \phi - 1}}. \tag{6.18}$$

Since Q is independent of the die angle, it is convenient to introduce the dimensionless strain rate intensity factor as

$$d = D \frac{\pi D_1^{5/2}}{Q} \tag{6.19}$$

Substituting Eq. (6.18) into Eq. (6.19) and eliminating the ratio B/Q by means of Eq. (6.11) yield

$$d = -\left(\frac{D_1}{r} \right)^{5/2} \sqrt{\frac{\tan \phi}{2 \, (c \tan \phi - 1)}} \left\{ \int_0^{\pi/2} \frac{\exp \left[-Q_1 \left(\gamma, \theta \right) \right] \sin \gamma \tan \theta \sin \theta}{\left[\cos \gamma + \left(2\sqrt{3} \sin \gamma - c \right) \tan \theta \right]} d\gamma \right\}^{-1},$$

$$Q_1 \left(\gamma, \theta \right) = 2\sqrt{3} \int_0^{\gamma} \frac{\cos \chi \tan \theta}{\cos \chi + \left(2\sqrt{3} \sin \chi - c \right) \tan \theta} d\chi \tag{6.20}$$

The distribution of the dimensionless strain rate intensity factor along the friction surface is depicted in Fig. 6.15. In this figure,

$$X = \frac{r}{D_1} - \frac{1}{2 \sin \phi} \tag{6.21}$$

Therefore, $X = 0$ corresponds to the outlet boundary independently of the value of ϕ. It is seen from Fig. 6.15 that the magnitude of the strain rate intensity factor increases with ϕ at the exit from the die. On the other hand, it is seen from the experimental results shown in Table 6.2 that the thickness of the fine grain layer also increases with ϕ when this angle varies in the range $5^0 \le \phi \le 15^0$. Therefore, there is a correlation between the strain rate intensity factor and the thickness of the fine grain layer generated near the friction surface.

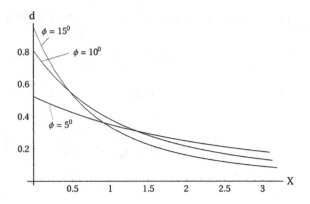

Fig. 6.15 Variation of the dimensionless strain rate intensity factor with X at several values of ϕ

6.2 Upper Bound Method

In utilizing the upper bound method, the velocity fields are typically approximated by a class of assumed functions. This is true even when finite element methods are used. However, as is known from other areas of mechanics, the assumed functions can have a large effect on the accuracy of the result. It is especially important to take into account the behaviour of the real functions that must exist near singular points, lines and surfaces. The singular behaviour of the velocity field presented in Chap. 2 has been already adopted in upper bound solutions. Most of these solutions deal with welded structures [1–4, 7, 9, 10, 15]. Few solutions have been provided for metal forming processes [12–14]. In the present chapter, the approach of using singular velocity fields in upper bound solutions is illustrated by means of a solution for hollow disk forging. The solution has been given in [11]. This solution is based on a new kinematically admissible velocity field that accounts for the singular behaviour of the velocity field near maximum friction surfaces. Other general kinematically admissible velocity fields for various forming processes have been proposed in [5, 20, 23, 24, 32, 33, 35]. The fields proposed in [32, 35] can be adopted to solve axisymmetric forging problems. The approach developed in [35] is to build up axisymmetric kinematically admissible velocity fields from regions in which the axial velocity is constant and the radial velocity is inversely proportional to the radius in a cylindrical coordinate system. The associated power dissipation has been calculated in this work as well. This approach has been adopted in [37] to study a radial forging process. A continuous velocity field satisfying the equation of incompressibility has been built up in [32]. The associated power dissipation has not been found. A possibility to introduce a rigid region near frictional interfaces has been mentioned, but no specific applications have been made. The velocity field proposed in [32] includes several other kinematically admissible velocity fields as special cases, for example those used in [17, 31]. In these works, the associated power dissipation has been calculated. Several upper bound solutions for axisymmetric ring forging have been

Fig. 6.16 Parameters in ring forging and coordinate system

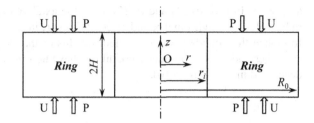

also found in [31]. Solutions based on the velocity field proposed in [17] have been derived in [18, 19]. Other upper bound solutions for axisymmetric ring forging have been given in [16, 30, 36].

6.2.1 Statement of the Problem

A ring of inner radius r_i, outer radius R_0 and thickness $2H$ is forged between two flat dies moving towards each other with velocity U. Cylindrical coordinates (r, θ, z) are taken, with the z axis taken as the axis of symmetry of the process. Also, the process is symmetric relative to the plane $z = 0$ (Eq. 6.16). The material of the ring is rigid perfectly plastic. The constitutive equations are given in Eqs. (1.18) and (1.19). Since $z = 0$ is a plane of symmetry for the flow, it is sufficient to consider the region $z \geq 0$. The dies are rough and the friction law is

$$\sigma_{rz} = \pm mk_0 \qquad (6.22)$$

at $z = H$. Here σ_{rz} is the shear stress in the cylindrical coordinates and m is the friction factor, $0 \leq m \leq 1$. Equation (6.22) is valid at sliding. At sticking, this equation must be replaced with the condition that the relative velocity between the ring and the die vanishes. The maximum friction law in the form of Eq. (1.28) is obtained from Eq. (6.22) if $m = 1$. By symmetry,

$$\sigma_{rz} = 0 \qquad (6.23)$$

and

$$u_z = 0 \qquad (6.24)$$

at $z = 0$. Moreover,

$$u_z = -U \qquad (6.25)$$

at $z = H$. In Eqs. (6.24) and (6.25), u_z is the axial velocity (u_r will stand for the radial velocity). Additional boundary conditions are in general required to completely formulate the boundary value problem for this or that ring forging process. However, these additional boundary conditions have no effect on the general approach developed in [11] (Fig. 6.16).

6.2.2 General Approach

In order to build up a kinematically admissible velocity field it is convenient to introduce the following dimensionless quantities:

$$h = \frac{H}{R_0}, \quad t = \frac{r_i}{R_0}, \quad \rho = \frac{r}{R_0}, \quad \zeta = \frac{z}{H}. \tag{6.26}$$

A typical assumption concerning the axial velocity is

$$\frac{u_z}{U} = -\zeta. \tag{6.27}$$

It is worthy of note that the through thickness distribution of the axial velocity is linear in the famous Prandtl's solution for the somewhat similar problem of compression of a block between rough plates in plane strain (see, for example, [25]). It is also known that this Prandtl's solution is a very good approximation to the accurate slip-line solution given in [26] when the thickness of the block is small enough. Therefore, it seems that Eq. (6.27) supplies a good approximation to the real velocity for thin rings. Using Eq. (6.26) the incompressibility equation in the cylindrical coordinates is written as

$$\frac{\partial u_r}{\partial \rho} + \frac{u_r}{\rho} + \frac{\partial u_z}{h \partial \zeta} = 0. \tag{6.28}$$

Substituting Eq. (6.27) into Eq. (6.28) and integrating yield

$$\frac{u_r}{U} = \frac{\rho}{2h} + \frac{f(\zeta)}{\rho}. \tag{6.29}$$

where $f(\zeta)$ is an arbitrary function of ζ. The velocity field shown in Eqs. (6.27) and (6.29) gives the following non-zero strain rate components in the cylindrical coordinates

$$\xi_{rr} = \frac{U}{H}\left[\frac{1}{2} - \frac{h}{\rho^2}f(\zeta)\right], \quad \xi_{\theta\theta} = \frac{U}{H}\left[\frac{1}{2} + \frac{h}{\rho^2}f(\zeta)\right], \tag{6.30}$$

$$\xi_{zz} = -\frac{U}{H}, \quad \xi_{rz} = \frac{Ug(\zeta)}{2H\rho}$$

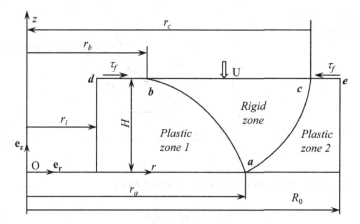

Fig. 6.17 Rigid and plastic zones in ring forging

where $g(\zeta) = df/d\zeta$. In the cylindrical coordinates, Eq. (1.11) becomes

$$\xi_{eq} = \sqrt{\frac{2}{3}}\sqrt{\xi_{rr}^2 + \xi_{\theta\theta}^2 + \xi_{zz}^2 + 2\xi_{rz}^2}. \qquad (6.31)$$

Substituting Eq. (6.30) into Eq. (6.31) results in

$$\xi_{eq} = \frac{U}{H\rho^2}\sqrt{\rho^4 + \frac{\rho^2 g^2(\zeta)}{3} + \frac{4}{3}h^2 f^2(\zeta)}. \qquad (6.32)$$

The presence of friction at $z = H$ demands $g(\zeta) \neq 0$ in exact solutions. In this case, the radial velocity shown in Eq. (6.29) is not compatible with the boundary condition $u_r = 0$ for $r = r_a = $ constant (or $\rho = \rho_a = r_a/R_0$ where $t \leq \rho_a \leq 1$), unless a rigid zone occurs in the vicinity of the surface $\rho = \rho_a$. On the other hand, the aforementioned boundary condition is often used to formulate the boundary value problem for ring forging processes [16, 17, 19, 31]. In the most general case, the rigid zone separates two plastic zones (Fig. 6.17). The motion of this rigid zone is a translation along the z axis with velocity U. The boundaries between the rigid and plastic zones (ab and ac in Fig. 6.17) are velocity discontinuity surfaces (lines in meridian planes). If $H/(R_0 - r_i)$ is small enough then it is reasonable to assume that the rigid zone penetrates the thickness of the ring. Therefore, the z coordinate of point a is zero. Consider a generic velocity discontinuity line. Let $\mathbf{e}_r$ and $\mathbf{e}_z$ be the unit base vectors of the cylindrical coordinate system. Then, the velocity vector in the rigid zone is given by

$$\mathbf{V_r} = -U\mathbf{e}_z. \qquad (6.33)$$

Fig. 6.18 Generic velocity
discontinuity line

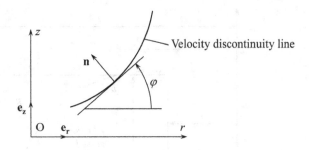

The velocity vector in the plastic zone is represented as

$$\mathbf{V_p} = u_r \mathbf{e_r} + u_z \mathbf{e_z}. \tag{6.34}$$

Here u_r and u_z are given by Eqs. (6.27) and (6.29). Let $\mathbf{n}$ be the unit normal vector
to the velocity discontinuity line and φ be the angle between the tangent to this line
and the r axis, measured from the axis anti-clockwise (Fig. 6.18). From geometry of
this figure

$$\mathbf{n} = -\sin\varphi\mathbf{e_r} + \cos\varphi\mathbf{e_z} \quad \text{and} \quad \tan\varphi = \frac{dz}{dr}. \tag{6.35}$$

The normal velocity must be continuous across the velocity discontinuity line.
Therefore, $\mathbf{V_r} \cdot \mathbf{n} = \mathbf{V_p} \cdot \mathbf{n}$ on this line. Substituting Eqs. (6.27), (6.29), (6.33), (6.34)
and (6.35) into this equation and using Eq. (6.26) lead to the following equation for
the velocity discontinuity line

$$\frac{d\rho}{d\zeta} = \frac{h}{(1-\zeta)}\left[\frac{\rho}{2h} + \frac{f(\zeta)}{\rho}\right]. \tag{6.36}$$

Using the substitution

$$\eta = \rho^2 \tag{6.37}$$

it is possible to reduce Eq. (6.36) to a linear differential equation. Its general solution is

$$\eta = \eta_d(\zeta) = \frac{2hF(\zeta) + C}{1-\zeta} \tag{6.38}$$

where C is constant of integration, the second equation defines $\eta_d(\zeta)$ and $F(\zeta) = \int_1^\zeta f(\chi)\,d\chi$. Here χ is a dummy variable of integration. It is seen from Eq. (6.38) that
$\eta_d(\zeta) \to \infty$ as $\zeta \to 1$, unless $C = 0$. Assume that the latter condition is satisfied.
Then, the right hand side of Eq. (6.38) reduces to the expression 0/0 at $\zeta = 1$.

Applying l'Hospital rule shows that $\eta_d(1) = -2hF'(1)$. Here $F'(1)$ denotes the derivative of the function $F(\zeta)$ at $\zeta = 1$. Using the definition for the function $F(\zeta)$ given after Eq. (6.38) it is possible to find that

$$\eta_0 = -2hf(1) \tag{6.39}$$

where η_0 is the value of η at $\zeta = 1$. Equation (6.38) becomes

$$\eta_d(\zeta) = \frac{2hF(\zeta)}{1-\zeta}. \tag{6.40}$$

The velocity discontinuity line passes through point a (Fig. 6.17) whose coordinates are $\zeta = 0$ and $\rho = \rho_a$. Therefore, it is seen from Eqs. (6.37) and (6.40) that

$$F = \frac{\rho_a^2}{2h} \tag{6.41}$$

at $\zeta = 0$.

The amount of velocity jump across the velocity discontinuity line is given by $[u_\tau] = |\mathbf{V_r} - \mathbf{V_p}|$. Then, it follows from Eqs. (6.27), (6.29), (6.33), and (6.34) that

$$[u_\tau] = U\sqrt{(1-\zeta)^2 + \left[\frac{\rho_d(\zeta)}{2h} + \frac{f(\zeta)}{\rho_d(\zeta)}\right]^2}. \tag{6.42}$$

where $\rho_d(\zeta)$ is determined from Eqs. (6.37) and (6.40) as

$$\rho_d(\zeta) = \sqrt{2h}\sqrt{\frac{F(\zeta)}{1-\zeta}}. \tag{6.43}$$

Using Eq. (6.26) an infinitesimal length element of the velocity discontinuity line is represented as

$$dL = \sqrt{\left(\frac{dr}{dz}\right)^2 + 1}\,dz = H\sqrt{h^{-2}\left(\frac{d\rho}{d\zeta}\right)^2 + 1}\,d\zeta.$$

Here the derivative $d\rho/d\zeta$ is understood to be taken at $\rho = \rho_d(\zeta)$. Therefore, this derivative can be replaced with the right hand side of (6.36) in which ρ is replaced with $\rho_d(\zeta)$ to give

$$dL = \frac{H}{(1-\zeta)}\sqrt{\left[\frac{\rho_d(\zeta)}{2h} + \frac{f(\zeta)}{\rho_d(\zeta)}\right]^2 + (1-\zeta)^2}\,d\zeta. \tag{6.44}$$

Having the kinematically admissible velocity field it is possible to calculate the plastic work rate involved in the upper bound theorem. In particular, the work rate in plastic zones is given by

$$W_V = 2\pi\sigma_0 \iint \xi_{eq} r\, dr\, dz. \tag{6.45}$$

Here σ_0 is the yield stress in tension. In the case of the von Mises yield criterion (see Eq. (1.18)) $\sigma_0 = \sqrt{3}k_0$.

Consider plastic zone 1 (Fig. 6.17) and denote the plastic work rate in this zone by $W_V^{(1)}$. Let $f(\zeta) = f_1(\zeta)$, $g(\zeta) = g_1(\zeta)$ and $F(\zeta) = F_1(\zeta)$ in this zone. Also, η_b will stand for η_0 and $\eta_{ab}(\zeta)$ for $\eta_d(\zeta)$. Therefore, the r coordinate of point b is $r_b = R_0\sqrt{\eta_b}$. Using Eqs. (6.26), (6.32), (6.37) and (6.45) yields

$$\frac{W_V^{(1)}}{\pi U \sigma_0 R_0^2} = \int_0^1 \int_{t^2}^{\eta_{ab}(\zeta)} \eta^{-1}\sqrt{\eta^2 + \frac{\eta}{3}g_1^2(\zeta) + \frac{4}{3}h^2 f_1^2(\zeta)}\, d\eta\, d\zeta \tag{6.46}$$

if $\eta_a \geq t^2$ and

$$\frac{W_V^{(1)}}{\pi U \sigma_0 R_0^2} = \int_0^{\zeta_1} \int_{t^2}^{\eta_{ab}(\zeta)} \eta^{-1}\sqrt{\eta^2 + \frac{\eta}{3}g_1^2(\zeta) + \frac{4}{3}h^2 f_1^2(\zeta)}\, d\eta\, d\zeta \tag{6.47}$$

if $\eta_b < t^2$. It is evident that the velocity discontinuity line ab (Fig. 6.17) and the line $\rho = t$ intersect in the range $0 < \zeta < 1$ if $\eta_b < t^2$. The $\zeta -$ coordinate of this point is ζ_1. It follows from Eqs. (6.37) and (6.40) that the value of ζ_1 is determined from the following equation

$$t^2 = \frac{2hF(\zeta_1)}{1 - \zeta_1}. \tag{6.48}$$

Consider plastic zone 2 (Fig. 6.17) and denote the plastic work rate in this zone by $W_V^{(2)}$. Let $f(\zeta) = f_2(\zeta)$, $g(\zeta) = g_2(\zeta)$ and $F(\zeta) = F_2(\zeta)$ in this zone. Also, η_c will stand for η_0 and $\eta_{ac}(\zeta)$ for $\eta_d(\zeta)$. Therefore, the r coordinate of point c is $r_c = R_0\sqrt{\eta_c}$. Using Eqs. (6.26), (6.32), (6.37) and (6.45) yields

$$\frac{W_V^{(2)}}{\pi U \sigma_0 R_0^2} = \int_0^1 \int_{\eta_{ac}(\zeta)}^1 \eta^{-1}\sqrt{\eta^2 + \frac{\eta}{3}g_2^2(\zeta) + \frac{4}{3}h^2 f_2^2(\zeta)}\, d\eta\, d\zeta \tag{6.49}$$

if $\eta_c \leq 1$ and

$$\frac{W_V^{(2)}}{\pi U \sigma_0 R_0^2} = \int\limits_{0}^{\zeta_2} \int\limits_{\eta_{ac}(\zeta)}^{1} \eta^{-1} \sqrt{\eta^2 + \frac{\eta}{3}g_2^2(\zeta) + \frac{4}{3}h^2 f_2^2(\zeta)} d\eta d\zeta \tag{6.50}$$

if $\eta_c > 1$. The velocity discontinuity line ac (Fig. 6.17) and the line $\rho = 1$ intersect in the range $0 < \zeta < 1$ if $\eta_c > 1$. The $\zeta-$ coordinate of this intersection point is ζ_2. It follows from Eqs. (6.37) and (6.40) that the value of ζ_2 is determined from the following equation

$$\frac{2hF(\zeta_2)}{1 - \zeta_2} = 1.$$

It is worthy of note that integration with respect to η in Eqs. (6.46), (6.47), (6.49) and (6.50) can be carried out in terms of elementary functions, which significantly facilitates further numerical integration with respect to ζ.

The plastic work rate at velocity discontinuity lines is determined as

$$W_d = \frac{2\pi \sigma_0}{\sqrt{3}} \int [u_\tau] r dL. \tag{6.51}$$

Consider velocity discontinuity line ab (Fig. 6.17) and denote the work rate at this line by $W_d^{(1)}$. Substituting Eqs. (6.42), (6.43) and (6.44) into Eq. (6.51) and using Eq. (6.26) result in

$$\frac{W_d^{(1)}}{\pi U \sigma_0 R_0^2} = \frac{2\sqrt{2}\sqrt{h}}{\sqrt{3}} \int\limits_{0}^{1} \frac{\sqrt{F_1(\zeta)} J_1(\zeta)}{(1 - \zeta)^{3/2}} d\zeta, \tag{6.52}$$

$$J_1(\zeta) = \frac{F_1(\zeta)}{2(1 - \zeta)} + f_1(\zeta) + \frac{(1 - \zeta)f_1^2(\zeta)}{2F_1(\zeta)} + h(1 - \zeta)^2$$

if $\eta_b \geq t^2$ and

$$\frac{W_d^{(1)}}{\pi U \sigma_0 R_0^2} = \frac{2\sqrt{2}\sqrt{h}}{\sqrt{3}} \int\limits_{0}^{\zeta_1} \frac{\sqrt{F_1(\zeta)} J_1(\zeta)}{(1 - \zeta)^{3/2}} d\zeta, \tag{6.53}$$

$$J_1(\zeta) = \frac{F_1(\zeta)}{2(1 - \zeta)} + f_1(\zeta) + \frac{(1 - \zeta)f_1^2(\zeta)}{2F_1(\zeta)} + h(1 - \zeta)^2$$

if $\eta_b < t^2$. Analogously, the plastic work rate at velocity discontinuity line ac (Fig. 6.17) denoted by $W_d^{(2)}$ is

$$\frac{W_d^{(2)}}{\pi U \sigma_0 R_0^2} = \frac{2\sqrt{2}\sqrt{h}}{\sqrt{3}} \int_0^1 \frac{\sqrt{F_2(\zeta)} J_2(\zeta)}{(1-\zeta)^{3/2}} d\zeta, \tag{6.54}$$

$$J_2(\zeta) = \frac{F_2(\zeta)}{2(1-\zeta)} + f_2(\zeta) + \frac{(1-\zeta) f_2^2(\zeta)}{2F_2(\zeta)} + h(1-\zeta)^2$$

if $\eta_c \leq 1$ and

$$\frac{W_d^{(2)}}{\pi U \sigma_0 R_0^2} = \frac{2\sqrt{2}\sqrt{h}}{\sqrt{3}} \int_0^{\zeta_1} \frac{\sqrt{F_2(\zeta)} J_2(\zeta)}{(1-\zeta)^{3/2}} d\zeta, \tag{6.55}$$

$$J_2(\zeta) = \frac{F_2(\zeta)}{2(1-\zeta)} + f_2(\zeta) + \frac{(1-\zeta) f_2^2(\zeta)}{2F_2(\zeta)} + h(1-\zeta)^2$$

if $\eta_c > 1$.

The work rate at the friction surface $z = H$ is

$$W_f = \frac{2\pi m \sigma_0}{\sqrt{3}} \int |u_r| r \, dr.$$

It is understood here that the radial velocity is to be calculated at $z = H$. Then, using Eqs. (6.26), (6.29), (6.37) and (6.39) gives

$$\frac{W_f^{(1)}}{\pi U \sigma_0 R_0^2} = \frac{m}{\sqrt{3}h} \int_t^{\sqrt{\eta_b}} \left(\eta_b - \rho^2 \right) d\rho \tag{6.56}$$

$$\frac{W_f^{(2)}}{\pi U \sigma_0 R_0^2} = \frac{m}{\sqrt{3}h} \int_{\sqrt{\eta_c}}^1 \left(\rho^2 - \eta_c \right) d\rho.$$

Here $W_f^{(1)}$ and $W_f^{(2)}$ are the work rates dissipated by friction at bd and ce, respectively (Fig. 6.17). Integrating in Eq. (6.56) gives

$$\frac{W_f^{(1)}}{\pi U \sigma_0 R_0^2} = \frac{m}{\sqrt{3}h} \left[\frac{2}{3} \eta_b \sqrt{\eta_b} + t \left(\frac{t^2}{3} - \eta_b \right) \right] \tag{6.57}$$

and

$$\frac{W_f^{(2)}}{\pi U \sigma_0 R_0^2} = \frac{m}{\sqrt{3}h} \left(\frac{2}{3} \eta_c \sqrt{\eta_c} - \eta_c + \frac{1}{3} \right). \tag{6.58}$$

Equations (6.57) and (6.58) are valid only if $\eta_b \geq t^2$ and $\eta_c \leq 1$, respectively. It is evident that $W_f^{(1)} = 0$ if $\eta_b \leq t^2$ and $W_f^{(2)} = 0$ if $\eta_c \geq 1$.

Having the plastic work rate it is possible to evaluate the force required to deform the ring by means of the upper bound theorem. In particular, it follows from this theorem that [25]

$$p_u = \frac{P_u}{\pi \sigma_0 R_0^2} = \frac{W_V^{(1)} + W_V^{(2)} + W_d^{(1)} + W_d^{(2)} + W_f^{(1)} + W_f^{(2)} + W_a}{\pi U \sigma_0 R_0^2}. \tag{6.59}$$

Here P_u is an upper bound on the force required to deform the ring, p_u is its dimensionless representation and W_a is the plastic work rate resulting from boundary conditions other than (6.22). As usual, the function $f(\zeta)$ can be represented by a combination of elementary functions. This combination may involve some arbitrary constants. These constants should be determined by minimizing the right hand side of Eq. (6.59). The real velocity field satisfies Eq. (2.20) if $m = 1$. In the case under consideration, the correct asymptotic behavior of the velocity field is obtained if

$$g(\zeta) = O\left(\frac{1}{\sqrt{1-\zeta}}\right) \tag{6.60}$$

as $\zeta \to 1$. The associated flow rule is given by Eq. (1.19). In the cylindrical coordinates, the equation of this rule for the shear strain rate becomes $\xi_{rz} = 3\lambda\sigma_{rz}$. It follows from this equation and Eq. (6.30) that the boundary condition (6.23) is satisfied if $f(\zeta)$ is an even function of its argument. Note that neither this property of the function $f(\zeta)$ nor Eq. (6.60) are necessary conditions. However, it is advantageous to take both into account since the behavior of such kinematically admissible velocity fields coincides with that of the corresponding exact velocity fields, even though the latter may be unknown.

6.2.3 Illustrative Example

As an illustrative example, forging inside a confined chamber with the size of the chamber equal to the outer radius of the ring is considered (Fig. 6.19). The radial velocity vanishes at $r = R_0$. Therefore, $\rho_a = 1$ (ρ_a has been introduced after Eq. (6.32)), plastic zone 2 (Fig. 6.17) does not exist and $W_V^{(2)} = W_d^{(2)} = W_f^{(2)} = 0$. The friction stress at $r = R_0$ is $m_i\sigma_0/\sqrt{3}$. Therefore,

$$W_a = \frac{2U\pi\sigma_0 R_0^2 h m_i}{\sqrt{3}}. \tag{6.61}$$

Fig. 6.19 Forging inside
confined chamber

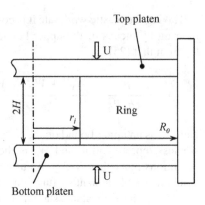

Substituting Eqs. (6.46), (6.47), (6.52), (6.53), (6.57), and (6.61) into Eq. (6.59)
yields

$$
p_u = \frac{P_u}{\pi \sigma_0 R_0^2} = \int_0^1 \int_{t^2}^{\eta_{ab}(\zeta)} \eta^{-1} \sqrt{\eta^2 + \frac{\eta}{3} g_1^2(\zeta) + \frac{4}{3} h^2 f_1^2(\zeta)} \, d\eta \, d\zeta +
$$

$$
\frac{2\sqrt{2}\sqrt{h}}{\sqrt{3}} \int_0^1 \frac{\sqrt{F_1(\zeta)} F_2(\zeta)}{(1-\zeta)^{3/2}} d\zeta + \tag{6.62}
$$

$$
\frac{m}{\sqrt{3}h} \left[\frac{2}{3} \eta_b \sqrt{\eta_b} + t \left(\frac{t^2}{3} - \eta_b \right) \right] + \frac{2hm_i}{\sqrt{3}},
$$

$$
F_2(\zeta) = \frac{F_1(\zeta)}{2(1-\zeta)} + f_1(\zeta) + \frac{(1-\zeta) f_1^2(\zeta)}{2F_1(\zeta)} + h(1-\zeta)^2.
$$

if $\eta_b \geq t^2$ and

$$
p_u = \frac{P_u}{\pi \sigma_0 R_0^2} = \int_0^{\zeta_1} \int_{t^2}^{\eta_{ab}(\zeta)} \eta^{-1} \sqrt{\eta^2 + \frac{\eta}{3} g_1^2(\zeta) + \frac{4}{3} h^2 f_1^2(\zeta)} \, d\eta \, d\zeta +
$$

$$
\frac{2\sqrt{2}\sqrt{h}}{\sqrt{3}} \int_0^{\zeta_1} \frac{\sqrt{F_1(\zeta)} F_2(\zeta)}{(1-\zeta)^{3/2}} \zeta + \frac{2hm_i}{\sqrt{3}}, \tag{6.63}
$$

$$
F_2(\zeta) = \frac{F_1(\zeta)}{2(1-\zeta)} + f_1(\zeta) + \frac{(1-\zeta) f_1^2(\zeta)}{2F_1(\zeta)} + h(1-\zeta)^2.
$$

if $\eta_b < t^2$.

One of the simplest ways to choose the function $f_1(\zeta)$ is to put

$$f_1(\zeta) = f_r(\zeta) = -\frac{\eta_b}{2h} - \beta\left(1 - \zeta^2\right) \tag{6.64}$$

where β is an arbitrary constant and the second equation defines $f_r(\zeta)$. It is evident that $f_r(\zeta)$ is an even function of ζ. Let $g_1(\zeta)$ be $g_r(\zeta)$ and $F_1(\zeta)$ be $F_r(\zeta)$ if $f_1(\zeta) = f_r(\zeta)$. Then, it follows from Eq. (6.64) that

$$g_r(\zeta) = 2\beta\zeta, \quad F_r(\zeta) = \frac{\eta_b}{2h}(1 - \zeta) + \frac{\beta}{3}\left(2 - 3\zeta + \zeta^3\right). \tag{6.65}$$

The function $g_r(\zeta)$ does not satisfy the condition (6.60). Equation (6.48) becomes

$$\left(t^2 - \eta_b\right)(1 - \zeta_1) = \frac{2h\beta}{3}\left(2 - 3\zeta_1 + \zeta_1^3\right). \tag{6.66}$$

The condition (6.41) is satisfied if

$$\beta = \frac{3(1 - \eta_b)}{4h}. \tag{6.67}$$

Eliminating β in Eqs. (6.64), (6.65) and (6.66) by means of Eq. (6.67) and substituting the resulting functions into Eqs. (6.62) and (6.63) determine p_u as a function of η_b. Minimizing this function with respect to η_b gives the best upper bound based on the assumption (6.64).

The simplest even function $f_1(\zeta)$ satisfying Eq. (6.60) is

$$f_1(\zeta) = f_a(\zeta) = -\frac{\eta_b}{2h} - \beta\sqrt{1 - \zeta^2}. \tag{6.68}$$

Here the second equation defines $f_a(\zeta)$. Let $g_1(\zeta)$ be $g_a(\zeta)$ and $F_1(\zeta)$ be $F_a(\zeta)$ if $f_1(\zeta) = f_a(\zeta)$. Then, it follows from Eq. (6.68) that

$$g_a(\zeta) = \frac{\beta\zeta}{\sqrt{1 - \zeta^2}}, \tag{6.69}$$

$$F_a(\zeta) = \frac{(1 - \zeta)\eta_b}{2h} + \frac{\beta}{2}\left(\arccos\zeta - \zeta\sqrt{1 - \zeta^2}\right).$$

Equation (6.48) becomes

$$\left(t^2 - \eta_b\right)(1 - \zeta_1) = \beta h\left(\arccos\zeta_1 - \zeta_1\sqrt{1 - \zeta_1^2}\right). \tag{6.70}$$

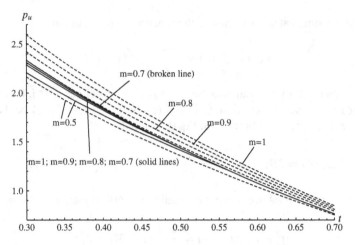

Fig. 6.20 Variation of the forging pressure with t at several values of m. Comparison of the present solution (*solid lines*) and the solution that does account for the shear strain rate (*broken lines*)

The condition (6.41) is satisfied if

$$\beta = \frac{2(1 - \eta_b)}{\pi h}. \tag{6.71}$$

As before, eliminating β in Eqs. (6.68), (6.69) and (6.70) by means of Eq. (6.71) and substituting the resulting functions into Eqs. (6.62) and (6.63) determine p_u as a function of η_b. Minimizing this function with respect to η_b gives the best upper bound based on the assumption (6.68).

Many upper bound solutions for axisymmetric forging processes are based on kinematically admissible velocity fields that neglect the shear strain rate in the cylindrical coordinate system [16–19]. For the sake of comparison such a solution is provided below assuming that $m_i = 0$. The kinematically admissible velocity field is obtained by putting $f(\zeta) = -1/(2h)$ in Eq. (6.29). Then, $g(\zeta) = 0$. The function $F(\zeta)$ is not required since there no velocity discontinuity line. Equation (6.59) reduces to

$$p_u = \frac{P_u}{\pi \sigma_0 R_0^2} = \frac{2 - \sqrt{1 + 3t^4}}{\sqrt{3}} + \tag{6.72}$$

$$\frac{1}{\sqrt{3}} \ln\left(\frac{1 + \sqrt{1 + 3t^4}}{3t^2}\right) + \frac{m}{\sqrt{3}h}\left[\frac{2}{3} + t\left(\frac{t^2}{3} - 1\right)\right].$$

The right hand side of Eqs. (6.62) and (6.63) has been minimized numerically assuming that $h = 0.25$. It is seen from these equations that the corresponding value of η_b is independent of m_i. Therefore, the numerical solution is illustrated for $m_i = 0$. In order to obtain p_u for any value of m_i, it is only necessary to add

$2hm_i/\sqrt{3}$. Both functions, $f_r(\zeta)$ and $f_a(\zeta)$, have been used. The variation of p_u found by means of the function $f_a(\zeta)$ with t at several values of m is depicted in Fig. 6.20 (solid lines). This solution supplies a better upper bound on the forging pressure than the solution based on the function $f_r(\zeta)$ when m is close to 1. However, the difference between the two solutions is very small ($< 3\%$) in the entire range of parameters investigated. Therefore, the other solution is not illustrated. The broken lines in Fig. 6.20 correspond to the solution (6.72). It is seen from this figure that the new solution is better than the solution (6.72) if $m > 0.6$ (approximately) and much better if m is close to 1. It is of importance that the new solution predicts no effect of m on p_u if t is large enough. This happens in the range $\eta_b \leq t^2$. In this case the regime of sticking occurs over the entire friction surface and the work rate dissipated by friction vanishes.

References

1. Alexandrov S (1999) A note on the limit load of welded joints of a rectangular cross section. Fat Fract Eng Mater Struct 22:449–452
2. Alexandrov S (2008) Limit load in bending of welded samples with a soft welded joint. J Appl Mech Techn Phys 49:340–345
3. Alexandrov SE (2012) Upper bound of the limit load for a tensile cylinder with a soft welded (soldered) joint containing a crack. J Appl Mech Techn Phys 53:793–799
4. Alexandrov S (2012) Upper bound limit load solutions for welded joints with cracks. Springer, Heidelberg
5. Abrinia K, Farahmand P, Parchami-Sarghin M (2014) Formulation of a new generalized kinematically admissible velocity field with a variable axial component for the forward extrusion of shaped sections. Int J Adv Manuf Technol 70:1427–1435
6. Alexandrov SE, Goldstein RV (2015) On constructing constitutive equations in metal thin layer near friction surfaces in material forming processes. Dokl Phys 60:39–41
7. Alexandrov SE, Goldstein RV, Tchikanova NN (1999) Upper bound limit load solutions for a round welded bar with an internal axisymmetric crack. Fat Fract Eng Mater Struct 22:775–780
8. Alexandrov S, Jeng Y-R, Hwang Y-M (2015) Generation of a fine grain layer in the vicinity of frictional interfaces in direct extrusion of AZ31 alloy. ASME J Manuf Sci Eng 137:Paper 051003
9. Alexandrov S, Kocak M (2007) Limit load analysis of strength undermatched welded T-joint under bending. Fat Fract Eng Mater Struct 30:351–355
10. Alexandrov S, Kocak M (2008) Effect of three-dimensional deformation on the limit load of highly weld strenght undermatched specimens under tension. Proc IMechE Part C J Mech Eng Sci 222:107–115
11. Alexandrov S, Lyamina E, Jeng J-R (2017) A general kinematically admissible velocity field for axisymmetric forging and its application to hollow disk forging. Int J Adv Manuf Technol 88:3113–3122
12. Alexandrov S, Mishuris G, Mishuris W, Sliwa RE (2001) On the dead zone formation and limit analysis in axially symmetric extrusion. Int J Mech Sci 43:367–379
13. Alexandrov S, Mustafa Y, Hwang Y-M, Lyamina E (2014) An accurate upper bound solution for plane strain extrusion through a wedge-shaped die. Sci World J 2014:Article 189070
14. Alexandrov S, Tzou G-Y, Hshia S-Y (2004) A new upper bound solution for a hollow cylinder subject to compression and twist. Proc IMechE Part C J Mech Eng Sci 218:369–375
15. Alexandrov S, Tzou G-Y, Hsia S-Y (2008) Effect of plastic anisotropy on the limit load of highly undermatched welded specimens in bending. Eng Fract Mech 75:3131–3140

16. Avitzur B, Sauerwine F (1978) Limit analysis of hollow disk forging. Part 1: upper bound. Trans ASME J Eng Ind 100:340–344
17. Avitzur B, Tyne CJV (1982) Ring formation: an upper bound approach. Part 1: flow pattern and calculation of power. Trans ASME J Eng Ind 104:231–237
18. Avitzur B, Tyne CJV (1982) Ring formation: an upper bound approach. Part 2: process analysis and characteristics. Thans ASME J Eng Ind 104:238–247
19. Avitzur B, Tyne CJV (1982) Ring formation: an upper bound approach. Part 3: constrained forging and deep drawing applications. Thans ASME J Eng Ind 104:248–252
20. Dogruoglu AN (2001) On constructing kinematically admissible velocity fields in cold sheet rolling. J Mater Process Technol 110:287–299
21. Griffiths BJ (1987) Mechanisms of white layer generation with reference to machining and deformation processes. ASME J Tribol 109:525–530
22. Goldstein RV, Alexandrov SE (2015) An approach to prediction of microstructure formation near friction surfaces at large plastic strains. Phys Mesomech 18:223–227
23. Gordon WA, Tyne JV, Moon YH (2007) Axisymmetric extrusion through adaptive dies part 1: flexible velocity fields and power terms. Int J Mech Sci 49:86–95
24. Haghighat H, Amjadian P (2011) A generalized velocity field for plane strain extrusion through arbitrarily curved dies. Trans ASME J Manuf Sci Eng 133:Paper 041006
25. Hill R (1950) The mathematical theory of plasticity. Clarendon Press, Oxford
26. Hill R, Lee EH, Tupper SJ (1951) A method of numerical analysis of plastic flow in plane strain and its application to the compression of a ductile material between rough plates. Trans ASME J Appl Mech 18:46–52
27. Hwang YM, Alexandrov S, Jeng YR, Huang TH, Naimark OB (2013) Finite element analyses of extrusion with a two stage die and manufacturing of gradient micro-structures. Key Eng Mater 528:23–31
28. Hwang YM, Chen Y, Alexandrov S (2016) Manufacture of magnesium tubes with gradient hardness distribution using a two-stage porthole extrusion die. Key Eng Mater 716:49–54
29. Hwang YM, Huang TH, Alexandrov S (2015) Manufacture of gradient microstructures of magnesium alloys using two - stage extrusion dies. Steel Res Int 86:956–961
30. Lee JH (1988) Upper Bound analysis of the upsetting of pressure sensitive polymeric rings. J Mater Process Technol 30:601–612
31. Liu JY (1971) Upper-bound solutions of some axisymmetric cold forging problems. Trans ASME J Eng Ind 93:1134–1144
32. Nagpal V (1974) General kinematically admissible velocity fields for some axisymmetric metal forming problems. Trans ASME J Eng Ind 96:1197–1201
33. Osakada K, Niimi Y (1975) A study on radial flow field for extrusion through conical dies. Int J Mech Sci 17:241–254
34. Shield RT (1955) Plastic flow in a converging conical channel. J Mech Phys Solids 3:246–258
35. Wilson WRD (1977) A simple upper-bound method for axisymmetric metal forming problems. Int J Mech Sci 19:103–112
36. Wu M-C, Yeh W-C (2007) Effect of natural boundary conditions on the upper-bound analysis of upset forging of ring and disks. Mater Des 28:1245–1256
37. Wu Y, Dong X (2015) Upper bound analysis of forging penetration in a radial forging process. Int J Mech Sci 103:1–8

Index

© The Author(s) 2018
S. Alexandrov, *Singular Solutions in Plasticity*, SpringerBriefs
in Continuum Mechanics, DOI 10.1007/978-981-10-5227-9

Printed in the United States
By Bookmasters